Fossils

A Hamlyn Colour Guide
Fossils

by **Rudolf Prokop**

Illustrated by
Vladimír Krb

Hamlyn
London • New York • Sydney • Toronto

Translated by Margot Schierlová
Graphic design by Miloš Lang
Designed and produced by Artia for
The Hamlyn Publishing Group Limited
London ● New York ● Sydney ● Toronto
Astronaut House, Feltham, Middlesex, England

Copyright © 1981 Artia, Prague

All Rights Reserved.
No part of this publication may be
reproduced or transmitted in any form or
by any means, electronic or mechanical
including photocopying, recording, or any
information storage and retrieval system,
without permission in writing from the
copyright owner.

ISBN 0 600 35370 2
Printed in Czechoslovakia
3/15/0551-01

CONTENTS

What is palaeontology? 6
What are fossils and how do they originate? 9
The chronological division of the Earth's past 13
Colour illustrations 15
Fossils tell a story 208
A few words on how ho collect fossils 211
The fossil collector's equipment 214
The preparation and conservation of fossils 215
The fossil collection 217
The identification and classification of fossils 218
Index 221

WHAT IS PALAEONTOLOGY?

Today it is quite difficult to give a concise and unequivocal definition of palaeontology. That is because palaeontology is a young science in the throes of explosive development, so its boundaries and links with other allied sciences are continuously growing wider and more intricate. In principle, palaeontology can be said to be the science of life and the presentation of life from past geological ages. It investigates the morphological remains of organisms and the traces of their activities. It cannot study the shapes and activities of organisms whose original forms have been lost, such as graphite, oil, asphalt and other organic substances finely dispersed in sedimentary rocks, or biogenic minerals and rocks, e.g. hydrocarbons and sulphides. These are covered by other branches of science.

Although palaeontology is a young science, its history is a wonderful survey of human mistakes, gropings and victories. Prehistoric man was not too interested in fossils; his main worry was where the next meal was coming from, and any fossils he happened to find were used as ornaments or perhaps as talismans. It is interesting to note that he was already acquainted with the now popular trilobites; at all events, a late neolithic necklace found in France, which was composed mainly of Cainozoic (Tertiary) snail shells, also comprised a handsome specimen of the trilobite *Paradoxides gracilis* from Bohemian middle Cambrian strata. We can assume, therefore, that a kind of barter or trade in some fossils existed in neolithic times.

Palaeontology was likewise unknown to antiquity, since it had no direct bearing on man himself — as botany had through medicinal herbs, or zoology through livestock and game — and, although people often found fossils, they did not pay them much attention. Only a few scholars studied fossils with special reference to the mystery of their origin. As many as 2,500 years ago, some Greek philosophers already had the right idea about the nature of fossils, regarding them as the remains of animals dating back to ages when the place where they were found was under water. However, alongside these, there were completely erroneous views, e.g. that fossils were the work of a creative force (*vis plastica*) present in the rocks, or that from mud and salt solutions the action of the moon and the stars created agglomerates and structures resembling animal bones and shells. Another curious view came from the famous philosopher Aristoteles (384—322 B.C.), who claimed that, through some unknown method of generation, slime and sludge could give rise to an organic body capable of life; and, as throughout the whole of the Middle Ages Aristoteles was the only acknowledged source of information on nature, his theory about fossils was accepted up to the middle of the eighteenth century. Correct explanations of the origin of fossils, irrespective of whether they came from ancient philosophers or from great minds of the Middle Ages (e.g. Leonardo da Vinci and Bernard Palissy, the celebrated French potter), were shrugged off, ridiculed or more often than not frowned upon by official opinion.

By the end of the seventeenth and beginning of the eighteenth century, it was already clear that fossils were not such a simple problem after all. Various forms of excavation kept bringing more to light, and many scholars, not content with just describing them, began to compare them with extant organisms. They were careful not to touch on the question of the origin of these remains, however, as the influence of current theories was still strong, and besides, the Church did not take kindly to such novelties. For instance, the famous French scholar Georges Buffon (1707—1788) came into conflict with the Sorbonne faculty of theology over his views on the development of species and the effect of the environment on their mutability, and was obliged to retract his theory.

In consequence, this was a time when palaeontological facts accumulated but remained unclassified and unevaluated. Palaeontology was not yet a science, and the importance of fossils for geological chronology (determination of the age of rock strata) and for the biological sciences was still unknown. Fossil collecting was more in the nature of a craze, and the majority of naturalists and collectors collected them rather as freaks of nature or as interesting or decorative objects.

Views on fossils did not alter until the end of the eighteenth and beginning of the nineteenth century when the natural sciences developed as a whole. On the one hand, the long forgotten theories of ancient scholars were resurrected, while on the other, contemporary research workers began to rely more and more on their own judgement and experience than on the secondhand views of the official authorities.

This was when the foundations of scientific palaeontology were laid and the general laws of palaeontology (which are basically still valid today) were formulated. The English geologist William Smith (1769—1839) discovered that individual geological layers had their own characteristic fossils, and that layers containing the same fossils were thus the same geological age. His 'Law of the Same Fossils' laid the foundations of biostratigraphy, a science which is still one of the fundamental components of geological research.

Georges Cuvier (1769—1832), the celebrated Parisian professor of comparative anatomy, formulated yet another important law — the law of correlations — which makes it possible, from the shape and structure of a single bone, to deduce the shape of the animal's other bones, or even of its entire skeleton.

An invaluable, though long unappreciated, contribution was made by another French scholar, Jean-Baptiste Lamarck (1744—1829), who derived from his studies general conclusions on the evolution of life on the Earth covering the evolution of its fauna from primitive to more highly organized animals and finally to man, the influence of the environment on the evolution of new species and disappearance of old ones, and also a series of further revolutionary discoveries which laid the foundations of whole branches and trends of modern palaeontology.

Joachim Barrande (1799—1883) was another outstanding figure in nineteenth century palaeontology. His gigantic treatise on the Primary fauna of Bohemia (*Système Silurien du Centre de la Bohème*), comprising 24 volumes and over 1,300 splendid lithographic plates, is still a standard work for modern revisional studies by palaeontologists everywhere.

The end of the nineteenth century and the twentieth century are characterized by the general development of palaeontology, However, it was found that no science could be isolated from the other sciences, and that the individual natural sciences are interwoven and complement each other, as seen from the very nature of the evolution of life and the tremendous diversity of nature as a whole.

This association with other branches of science, particularly geology, led to arguments as to where modern palaeontology actually belonged, whether to geology or to biology. Today, it is considered to be a biological science; it can be concisely described as historical biology — as against the biological sciences dealing with extant nature — but complicated by the necessity of taking the time factor into account. When we bear in mind that palaeontology, unlike contemporary biology, is concerned with animals and plants which succeeded each other on the Earth over a period of almost 3.5 billion years, we can see that it is by no means easy to define.

The controversy as to whether palaeontology is a biological or a geological science stems from the fact that it has been closely associated with geology right from the beginning. That is only natural, since without a knowledge of geological processes it would be impossible to describe either the past conditions of life or the factors which have influenced its evolution up to the present day. Conversely, the geological sciences in general and geological practice in particular would not have progressed far without extensive use of the results achieved by palaeontology.

Modern research shows that this basic link is not sufficient. There are other branches of science with which palaeontology (or palaeobiology) works in close co-operation, such as geophysics, geography, oceanography, ecology (the science of the interrelationships between organisms and their environment) and many others. There is thus a world of difference between the present-day palaeontologist, with his modern methods and wide fields of interest, and his worthy, bewigged predecessors who paved the way for him by sorting out, describing and classifying the fossils.

WHAT ARE FOSSILS AND HOW DO THEY ORIGINATE?

Fossils are the remains, impressions or traces of the bodies of organisms which have been preserved in rocks from previous geological ages to the present day. Their structure must furnish distinct evidence that they were once living individuals and must give some idea of their appearance and construction; they must be covered with sediment and must date back to prehistoric times.

Genuine fossils, i.e. unmodified or only slightly modified remains of skeletons or parts of skeletons (teeth, bones, shells) are comparatively rare. In very old rocks which have undergone many changes over millions of years they are extremely rare. There are exceptions, however. For instance, in unconsolidated sand in the Leningrad region, there are free-lying brachiopod shells of the genus *Obolus* composed of calcium phosphate, which have kept their natural colour and lustre. The sand is by no means recent, since it was deposited in Ordovician seas some 450 million years ago. Among the most popular genuine fossils are the carcasses of mammals frozen in the eternal ice of Siberia and the skeletons of vertebrates fossilized in asphalt deposits in the Rancho la Brea region in California.

The best known types of fossils, in the broadest sense of the term, are those in which any remains of the original skeleton were lost during fossilization, so that all that is left in the rock is an almost complete impression of its internal or external structure. Exoskeletal materials are often dissolved by chemical processes and are replaced by minerals such as calcite, silica, pyrite, etc. which generally form a perfect replica of the original. Internal moulds are also common. These are formed when soft sediment enters the body cavity after the soft parts have decomposed and, in time becomes lithified, retaining the morphological details of the inner surface of the animal's exoskeleton. When the exoskeletal materials have been dissolved, the outer contours are preserved in the form of an impression. It is thus important to collect both parts, i.e. the core (internal mould) and the impression (external mould), or, as they are sometimes called, the positive and the negative.

As well as the actual remains, moulds or casts of organic skeletons, we today include among fossils evidence of activities that reflect aspects of the lives of fossil animals, such as traces produced when crawling, burrowing, feeding, resting, etc. These characteristic traces sometimes reveal the presence of animals which have not been preserved in any other form. For example, the edges of leaves in Mesozoic clays often have semicircular pieces bitten out of them. This is typical of ants of the genus *Atta*, which still abound in tropical forests, and, although the clays contain no remains of the Mesozoic ants themselves, the typically damaged leaves are reliable evidence that they existed during that era.

Structures of inorganic origin are not regarded as fossil, however. They include, for example, soft sediment structures, dendritic aggregates of iron and manganese oxides on stratal fractures (which the layman often takes to be fossilized moss), the filling of mud fissures and, of course, all artefacts, i.e. objects fashioned by man (e.g. flint implements), which are the province of the archaeologist.

Thus, we know what fossils are and roughly how they originate. Not all organic remains are capable of fossilization, however; and the chances that a dead animal's body or a dead plant will be preserved are slim, as everybody can verify simply by observation. Dead animals and plants either act as food for other organisms, or they decompose and disintegrate. Hard parts — bones, teeth, shells, wood and hard-cased fruits — have better chances of being preserved. Organisms composed only of soft tissues, like medusas, certain worms and flowers, etc., or soft body tissues are preserved only under exceptionally favourable conditions; as a rule, both the soft body and the seemingly tougher parts all disappear relatively soon. On land, dead bodies do not decompose everywhere at the same rate: in polar regions, the process is slower and in the tropics faster. In the sea, most dead bodies are quickly destroyed by bacteria, other animals, chemical processes or mechanical factors, e.g. the movement of the water.

Even if all the above pitfalls have been overcome, there are still a number of other conditions which have to be fulfilled before organic remains can be fossilized.

The body must be covered as quickly as possible with sediment, which protects it from destruction by atmospheric or hydrospheric factors, from biological decomposition or from mechanical damage. Protective deposits are more commonly formed in water than on dry

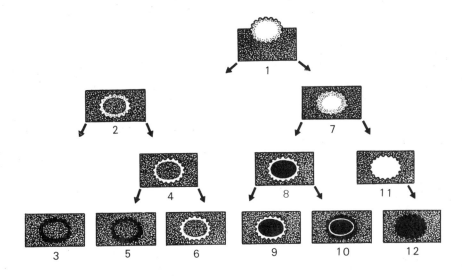

Fig. 1 presents in diagrammatic form the various ways in which a body case can be fossilized: 1 — a case partly buried in sediment; 2 — a case completely buried and filled with sediment; 3 — the matter of the case is gradually replaced by another material, thus producing an external cast and an inner mould; 4 — the case matter is dissolved; 5 — the original case is replaced by some other material, giving rise to an external cast and an inner mould; 6 — the case leaves a free space, with formation of a free inner mould and an external mould; 7 — the case is completely buried in sediment, but the inner cavity is empty; 8 — the inner cavity is secondarily infilled; 9 — the case matter is dissolved, producing a free inner mould and an external mould; 10 — the case matter is replaced by secondary material, with the resultant production of an external cast and an inner mould; 11 — the case matter disappears, leaving only an impression (external mould); 12 — the whole space left by the case and its cavity is filled with secondary material.

N. B. densely dotted areas = sediment; loosely dotted areas = original case material; black areas = secondary fossilization material (mineral or rock).

land, and fossilized aquatic animals are, therefore, found more often than terrestrial animals. The occurrence of these palaeontologically advantageous conditions on dry land is comparatively rare: only events like sandstorms, floods and volcanic activity producing fine ash provide sufficient material for the quick burial and subsequent fossilization of dead terrestrial animals and plants.

The petrographic composition of the sediment covering organic remains is very important for the formation and preservation of fossils. Fine-grained, impermeable sediments are always better than coarse-grained, permeable ones.

The chemistry of the covering sediment and its relationship to the chemical composition of the organism also need to match. For instance, the chalky shells of molluscs dissolve very quickly if they are immersed in acidic water or covered by mud containing hydrogen sulphide. Conversely, in lime-rich sediments, the chemistry of the environment and of the organism suit each other, and the shells have a better chance of preservation.

Natural resistance of a dead individual's body to destructive factors plays a major role. An elephant or rhinoceros naturally has a much better chance of preservation and fossilization than a tiny frog or mouse. Similarly, massive shells are preserved better than fragile, thin-walled shells, and wooden branches and trunks better than soft fruits. As already mentioned, animals and plants with no internal or external skeleton are unlikely to be fossilized.

Other factors facilitating the formation of fossils are pressure (generally the weight of the overlying rocks), temperature (which must not be so high as to destroy organic remains) and suitability of the mineral solutions which circulate in the rocks and eventually replace the original tissues of the organic skeleton. The commonest and best are aqueous calcium carbonate and silicic acid solutions.

The final essential is time, so that all the above factors and processes are able to take effect. The outcome of all these complicated processes and factors is the production of a fossil. That is still not the end, however. Fossils, especially those dating back to the earliest geological periods, lie in rocks which were subject to various modifications during the gigantic changes which took place in the Earth's crust. Rocks were folded, fractured, compressed and melted, and their structure was often considerably altered. Simultaneously the enclosed fossils were altered: they were rolled out, stretched, recrystallized and sometimes destroyed completely.

Obviously, therefore, fossils must be regarded as rare natural objects which have been preserved only as a result of the coincidence of exceptionally favourable circumstances. Every fossil is actually a unique record of life and its environment in past geological ages and of the ways in which life evolved. It is a record which can never be reproduced, and, if it is destroyed, its message dies with it. It would be like destroying a book of which only one copy existed and which could never be written. What does it matter that we are unable to read (and do not even understand) everything that is written in these fossil 'books'? Perhaps one day we shall really know all about them.

THE CHRONOLOGICAL DIVISION OF THE EARTH'S PAST

To be able to understand the incidence and evolution of associated groups of animals in relation to the general evolution of our planet, it is necessary to place the whole of the key Palaeozoic era in its correct chronological and factual context with the other geological eras. The basic stratigraphic table shows the various phases of the evolution of life on the Earth.

Era	Period	Duration in millions of years	Age in millions of years
Quaternary	Holocene	2	
	Pleistocene		2
Cainozoic (Tertiary)	Pliocene Miocene Oligocene Eocene Palaeocene	63	65
Mesozoic	Cretaceous Jurassic Triassic	160	225
Palaeozoic	Permian Carboniferous Devonian Silurian Ordovician Cambrian	345	570
Proterozoic		1,000	1,570
Archaic		?	4,000

As regards the Palaeozoic era itself, its simple division into early and late has long ceased to be sufficient. Its various periods have their own names, which are mostly taken from the regions where strata of the relevant ages appear in their classic form. In early Palaeozoic

13

strata we differentiate (from the oldest to the most recent) Cambrian, Ordovician, Silurian and Devonian formations, and in late Palaeozoic strata Carboniferous and Permian formations. According to the stage of evolution of the dominant groups of animals, the early Palaeozoic can be regarded as the age of invertebrates and fishes, and the late Palaeozoic as the age of amphibians and reptiles. The following table is a guide to the division of the whole era.

Era	Period		Duration in millions of years	Age in millions of years
Late Palaeozoic	Permian		55	280
	USA: Pennsylvanian	Europe:	45	
	....	Carboniferous	....	325 ....
	Mississippian		20	345
Early Palaeozoic	Devonian		50	395
	Silurian		35	430
	Ordovician		70	500
	Cambrian		70	570

In conclusion we must point out that in the descriptions and data on the various groups of animals of the prolific Palaeozoic era, the use of basic scientific morphological terms was unavoidable. Also it is possible that some of the names of the illustrated fossils are no longer valid and that they have been superseded. Palaeontology is developing so fast that the way in which individual species and genera are interpreted sometimes changes quickly too. Although some such changes may have been missed in this popular scientific manual, the actual purpose of the book is not affected.

Every one of the given species is named at the top of the page on the left; on the right is the name of the phylum (the primary division), with the name of the determinant lower systematic group below it.

COLOUR ILLUSTRATIONS

Fusulina sp.

Protozoa
Rhizopodea

One of the vast group of single-celled organisms (protozoans) which have inhabited every type of environment on our planet since time immemorial is the order Foraminiferida. Foraminiferids reached the peak of their evolution in the Mesozoic, and finds from this era are of considerable practical significance, as they can be utilized in oil prospecting. Nor were Palaeozoic species negligible. From the aspect of palaeontology, the only types of importance are those which formed a hard test of calcareous or siliceous matter, or by the agglutination (adhesion) of particles of inorganic matter. Marine foraminiferids form a substantial part of plankton (microscopic animals floating in the water), but some types crawl over the ocean bed (vagile benthos) or live attached to it (sessile benthos). The size of the tests varies from microscopic to several centimetres in length, and their shapes vary enormously.

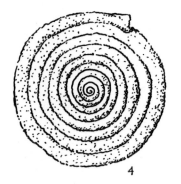

The commonest foraminiferids of the late Palaeozoic were members of the genus *Fusulina* (1), whose tests were coiled in an intricate, roomy spiral. In rocks they often weather out in the form of spindle-shaped grains (2). Fusuline 'chalk' often forms huge deposits in Carboniferous strata, all over the world.
In deposits washed out from weathered Palaeozoic rocks there are often fragments or even whole, simply branched tests up to 2 cm long, without a central chamber. They are foraminiferids of the sessile genus *Rhabdammina*, which already lived during the Silurian period and still inhabit the seas today (3). Only their environment has changed. Palaeozoic species lived mainly in shallow littoral water, whereas recent species chiefly inhabit the deep part of oceans in the arctic and the antarctic belt.

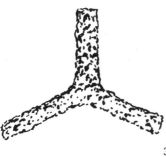

3

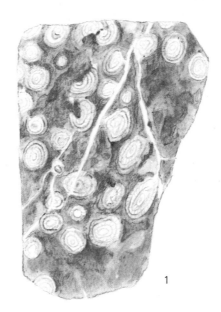

1

Many species of the ancient genus *Ammodiscus* have attractive and interestingly constructed tests coiled in a tight spiral (4). The surface of the test, which is relatively small (5 mm in diameter), is reinforced with fine sand (agglutinated test). The oldest species, which date back to the early Palaeozoic (Silurian) lived mainly in shallow water. Recent species frequent the deep, colder parts of the great oceans, where they drift in the water just above the sea bed.

2

Hydnoceras tuberosum

Porifera
Hyalospongea

Most people are acquainted with sea sponges — or at least with their horny skeleton in the form of the bath sponge. Few, however, are aware that they are holding the remains of an animal whose geological history goes back to the earliest Palaeozoic periods and possibly further still. Sponges (Porifera) are the most primitive multicellular animals; their body has not yet differentiated into tissues and organs. The usually pouch-like body has an open digestive cavity (paragaster) and walls reinforced with horny, chalky or siliceous spicules. The majority of sponges are sea-dwellers, and only a few live in fresh water. They are to be found at a wide range of depths and are distributed throughout all the oceans. Adult sponges are sessile and often change their shape to fit the base to which they are attached or their surroundings. Their height varies from a few millimetres to 1.5 m. The most important for palaeontological research are the calcareous sponges (Calcispongea) and particularly the siliceous sponges (Hyalospongea), since the material of which their skeletons, or at least the spicules, is composed is durable enough to be capable of fossilization.

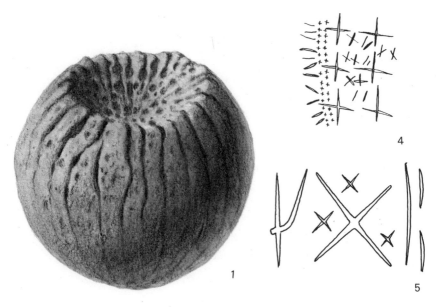

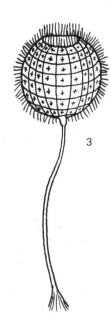

The best known siliceous Palaeozoic sponges include the members of the genus *Astylospongia*, whose spherical skeletons are composed of massive spicules fused together to form a strong network. The surface of the thick walls of *Astylospongia* is covered with regular ribs and grooves leading from the large oscule (excretory orifice) at the apex of the sponge to its base. These sponges are characteristic of Ordovician and Silurian rocks all over the world. The typical species *Astylospongia praemorsa* (1) abounds in German Silurian strata.

The members of the genus *Hydnoceras*, e.g. *Hydnoceras tuberosum* (2), were constructed on somewhat different lines. Whole skeletons of these sponges are known from upper Devonian and Carboniferous rocks in the USA and France. They are cornet-shaped, and their thin walls are prolonged into virtually regular rows of bulbous processes. The walls form a regular lattice composed of fine, simple and relatively long spicules.

Protospongia mononema (3, 4) and *Protospongia novaki* (5) are lower Palaeozoic sponges with free, unconnected spicules in their body wall. The siliceous spicules are often found scattered in rocks dating from early Cambrian to Ordovician in Europe, North America and southeastern Asia.

Cyclomedusa plana

Coelenterata
Hydrozoa

With their soft, bell-shaped bodies composed of a jelly-like tissue, medusas have little chance of reaching posterity as fossils. Such remains are therefore rare, although some are known from strata as old as 700 million years. They are obviously not genuine fossils but the impressions of medusa umbrellas on the surface of fine sediments or casts of the middle of the digestive cavity. The oldest known impressions of evolutionally primitive medusas were found in Proterozoic strata in southeastern Australia, at the famous Ediacara site, and similar finds were made subsequently in other countries. They all seem to have been individuals washed up by the sea on to a flat, muddy shore. The impressions of these medusas in the mud are often so distinct as to show traces of the mouth, the digestive cavity and the ring of tentacles on the under side of the umbrella. In Palaeozoic sediments, medusa impressions and casts are known chiefly from the North American Cambrian and the north European Ordovician, although here again, only isolated remains have been found. Nevertheless, there are sufficient of them to demonstrate the wealth of life in Palaeozoic seas and to enable us to document these animals' history.

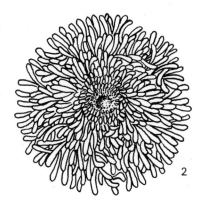

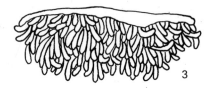

One of the best known Proterozoic medusa genera is *Cyclomedusa,* which had a concentrically grooved umbrella up to 20 cm across and a clearly discernible, circular central mouth. Some species are known from Australia and one, *Cyclomedusa plana* (1), also occurs in the USSR, in the Ukraine.

1

Eoporpita medusa had a remarkably constructed form with a flat, almost disc-like body, a small oral orifice and a dense wreath of tentacles covering practically the whole under side of the umbrella (2; 3 — side view). Finds made in fine Ediacara Precambrian sediments made it possible not only to reconstruct the body structure of this species in detail, but also to make plastic models of the animal as a whole. The diagrammatic cross section of the umbrella in Fig. 4 shows the elongate digestive cavity and the tentacles distributed over the whole under side of the umbrella. It was further found that the 700 million-year-old *Eoporpita* was a close relative of the still extant medusa genus *Porpita*, making it not only something of a scientific sensation, but also an 'alive' fossil.

4

Anaconularia anomala

Coelenterata
Scyphozoa

While true medusas, with their soft body practically incapable of fossilization, are exceptional in palaeontological research, the members of the subclass Conulata, the conulariids, are very important, particularly for the Palaeozoic. These exclusively marine animals had a thin, but elastic, flexible body wall composed of a substance resembling chitin and containing a small amount of calcium phosphate. The 'shell' was shaped mostly like a tapering pyramid with slightly bulging walls; its surface was adorned with transverse grooves and ridges, often combined with longitudinal sculpturing. The life of conulariids was also remarkable. Apart from a few small species or young individuals, which attached themselves to fixed or floating objects, the majority of conulariids could evidently swim. Their remains are to be found, therefore, in the most diverse types of sediments all over the world, from the Cambrian to the lower Triassic. At the end of the Palaeozoic, they suddenly began to die out and their last representatives disappeared in the Triassic period.

3

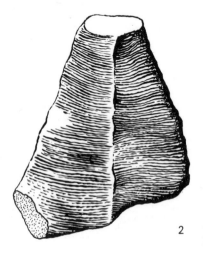

One of the first scientists to study conulariids was Joachim Barrande. He was particularly interested in the smooth, slim, tapering 'shells' of the species *Anaconularia anomala* (1), which are found in middle Ordovician quartzites in central Bohemia. Similar species, (e.g. *Conularia niagarensis*, whose 'shells' (2) had an interesting surface structure (3)), were later discovered in different strata of the same age in France, Turkey and North America, for instance. It can be concluded from their incidence that these conulariids probably lived in the shallow parts of the sea.

2

Other marine organisms considered to be related to the conulariids were characterized by a somewhat different type of 'shell' and evidently by a different mode of life as well. They had a smooth, slender, conical or dart-like, slightly curved chitin-phosphate 'shell', which sometimes had a small suction disc attached to the apex of the cone. As long ago as 1847, J. Hall described them as *Sphenothallus* (4) and they, or similar genera, are known from the most diverse Ordovician and Silurian sediments all over the world. Whole colonies of sphenothalli, attached by their discs to some floating or drifting body, e.g. beds of seaweed, have been discovered. In this way, the colonies were transported for long distances by the oceanic currents. Fig. 5 shows *Sphenothallus angustifolius.*

Entelophyllum prosperum

Coelenterata
Anthozoa

The corals (Anthozoa) are a large group of exclusively marine animals with stings and living either singly or in colonies. Their hard chalky or horny skeletons (known as coral) are excellent material for fossilization and often play a rock-forming role.

Corals of the order Rugosa occurred only in Palaeozoic seas, but they were distributed over the whole of the globe. They did not build atolls like many recent corals do, but generally formed dense growths on a 'not muddy' bed in shallow water with strong currents. Their remains are therefore to be found more in limestones and less in muddy or tuffaceous sediments. The calcareous skeletons of these corals, known as corallites, are usually found separately, and they have diverse forms: conical, cylindrical, saucer-shaped or goblet-shaped. It was rare for the corallites to unite and form clumps (colonies). The inner body cavity of a corallite is usually divided by radially organized, longitudinal ribs or septa. Horizontal plates (tabulae) are less strongly developed or even absent.

Entelophyllum, sometimes misnamed *Xylodes*, is a typical and prolific member of the rugose corals. Numerous species of the genus abound in Silurian limestones in Europe, North America, Asia and Australia. Their large cylindrical corallites, which have a deep body cavity and fine, dense septa, occur singly. *Entelophyllum prosperum* (1) formed growths on the bed of the Silurian sea round volcanic islands in central Bohemia.

Keriophyllum tabulatum (2) was another, this time colony-forming, species of rugose coral. These and similar corals formed round, cushion-like clumps of varying sizes on the bed of European middle Devonian seas. Younger growth stages were interpolated with the corallites of older individuals, so that in one colony we often find individuals of different shapes and sizes.

The rugose corals included some curious species, such as *Calceola sandalina*, known from Devonian strata practically all over the world. Its specific name, i.e. *sandalina*, is very apposite, since its deep cup (3) was shaped like the toe of a sandal and, in addition, was closed by a flat lid or operculum (4). The septa are reduced and appear only as tiny teeth on the dorsal wall of the corallite, where they dovetailed with teeth on the dorsal margin of the operculum; the whole formed a locking device allowing an opening and closing movement of the operculum. The unmistakable form of *Calceola* and their wide geographical distribution make them excellent index fossils for dating and correlating Devonian strata over large areas.

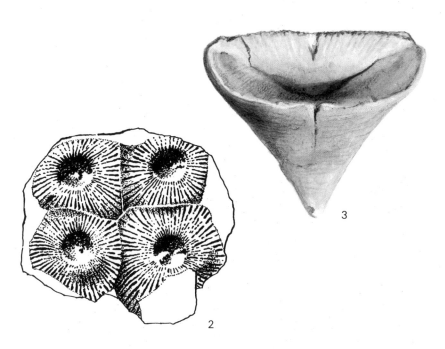

Favosites tachlowitzensis

Coelenterata
Anthozoa

Like rugose corals, the chalky skeletons of tabulate corals (order Tabulata) are found all over the world, but only in Palaeozoic marine sediments. The two groups had the same mode of life, and they often occur together. Tabulate corals are usually easy to distinguish. They frequently form massive colonies, which may measure over 1 m in diameter. The colonies consist of a large number of individuals all clustered together; they are usually bulbous or shaped like a round loaf and often send out trailers or branches. The vertical septa typical of rugose corals are missing or vestigial in tabulate corals, but there are well-developed and often numerous horizontal plates (tabulae). Tabulate corals are known chiefly from early Palaeozoic rocks; there are far fewer in Carboniferous and Permian strata. At the end of the Palaeozoic they disappeared, to make room, like Rugosa, for the more modern types of corals still encountered in the seas today.

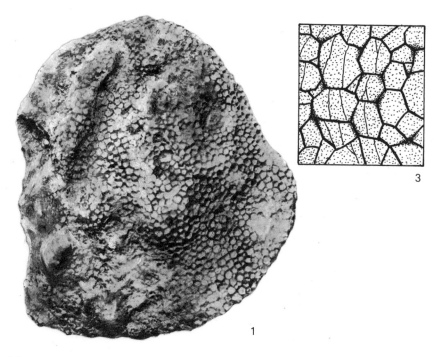

The best-known member of one of the families of tabulate corals is the genus *Favosites*, whose long, prismatic corallites are packed so closely together that their cups form a polygonal mosaic on the surface of the colony. The colonies, which have a honeycomb appearance and often weather out of the rocks, attracted the attention of scientists and collectors a long time ago, and the typical genus was already described by J. B. L. Lamarck in 1816. *Favosites tachlowitzensis* (1) is known from European Silurian tuffaceous limestones, while similar species abound in strata of the same age in the USA, southeastern Asia and Australia, for example.

Halysites catenularia (2) ('chain' coral), a representative of a further family of tabulate corals, is likewise very typical and easily distinguished. Its colonies are composed of long corallites fused laterally in long rows and forming characteristic chains in cross section. This species was described by Linnaeus as early as 1767 and is abundant in Silurian limestones throughout the world.

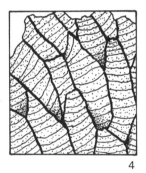

The cross and longitudinal sections of a colony of *Favosites gothlandicus* show the polygonal shape of the corallites (3) and the well-developed horizontal plates or tabulae (4).

Heliolites decipiens

Coelenterata
Anthozoa

The family Heliolitidae is a characteristic and easily distinguishable group of corals which, together with tabulate and rugose corals, formed thick growths on the well-lit beds of shallow Palaeozoic seas or grew in the more peaceful environment of coral reefs. They were likewise colonial corals. Their bulbous or hemispherical colonies can be identified from the rounded, rosette-like or stellate cross section of the individual corallites, which were relatively large and were separated from each other by a fine tubular network. The corallites had well-developed horizontal plates (tabulae) and fine vertical septa. Heliolitids are distributed all over the world, but they lived only in the early Palaeozoic, from the middle Ordovician to the middle Devonian.

3

4

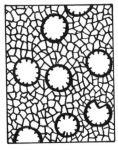

Heliolites decipiens (1; 2 — cross section) is a typical heliolitid coral known chiefly from European and North American Silurian strata. In Bohemian Palaeozoic rocks it is one of the commonest fossils in middle Silurian tuffaceous layers. These corals were thus not typical reef corals, but, together with other species, inhabited the beds of Silurian seas close to the shore, where the shallow water was well oxygenated by currents and repetitious waves.

Helioplasma, known from European and Asian Devonian layers, is another characteristic heliolitid genus. Sections of a colony of the species *Helioplasma kolihai* (3 — cross section) and *Helioplasma aliena* (4 — longitudinal section) show the typical structure of the corallites. As distinct from the preceding species, *Helioplasma* corallites are clustered together, have pronounced vertical septa (cross sections of the corallites are stellate) and less of the marginal network between the corallites. The characteristic genus *Heliolites* had relatively large, rounded corallites with minute septa; they were distributed comparatively loosely in the colony. Differences between the two typical genera concerned the mode of life as well as the morphology of the colonies. Unlike the Silurian *Heliolites,* the Devonian *Helioplasma* lived in the deep, still water round reef bodies.

Batostoma poctai

Bryozoa
Trepostomata

'Moss animals' (Bryozoa) are a very old animal phylum. They are known from Cambrian strata and are still an important component of marine faunas; in fresh water they are less common. 'Moss animals' are small creatures forming calcareous or chitinous colonies (zoaria) composed of the interconnected body cases (zooeciae) of large numbers of individuals. The colonies measure up to 70 cm across and have various forms, e.g. bulbous, hemispherical, tabulate, branched or reticulate. Together with corals, they often formed dense growths or even small reefs on shallow parts of current-swept sea beds. Trepostomatous 'moss animals' are known only from the Palaeozoic era; they were most prolific during the Ordovician. Their massive zoaria were composed of chalky tubes, often with transverse plates (diaphragms), adhering closely to each other and so resembling tabulate corals in appearance. The study of trepostomatous 'moss animals' and of bryozoans in general is quite complicated, and to determine a species often entails studying the intricate structure of the zoaria by means of thin sections. As a result, this is not a popular group with collectors.

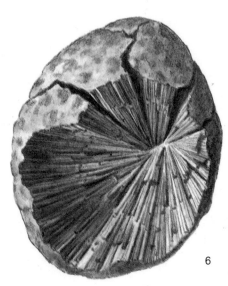

6

The important Ordovician genus *Batostoma*, which had dendritic, richly branching zoaria, was one of the commonest genera of the bryozoan order Trepostomata. Unlike most of the other genera, *Batostoma* frequented the quieter parts of shallow creeks. Their remains are usually found, therefore, in sandy-argillaceous-calcareous sediments, where they sometimes form significant horizons. *Batostoma poctai* (1) is an important fossil for the late Ordovician of central and southwestern Europe. A similar species, *Batostoma minnesotense* (2 — cross section of zoarium; 3 — longitudinal section of zoarium), is known from Ordovician strata in the USA.

Monotrypa undulata is a representative of another genus of trepostomatous bryozoans. It differs from the preceding species in respect of its massive, discoid or hemispherical colonies (zoaria) composed of prismatic

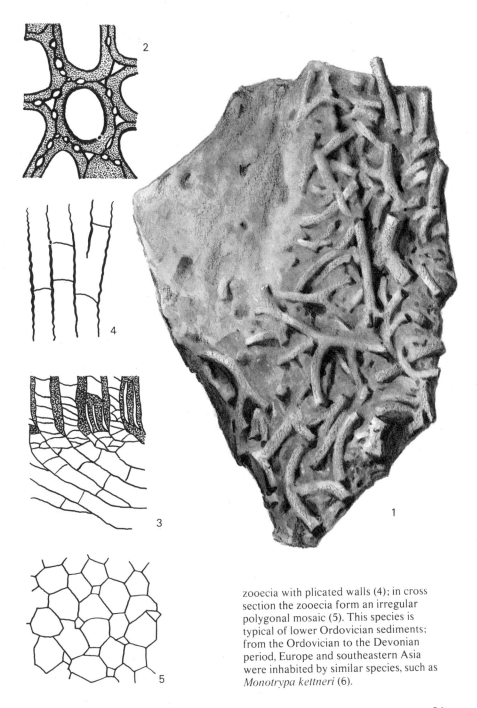

zooecia with plicated walls (4); in cross section the zooecia form an irregular polygonal mosaic (5). This species is typical of lower Ordovician sediments; from the Ordovician to the Devonian period, Europe and southeastern Asia were inhabited by similar species, such as *Monotrypa kettneri* (6).

Fenestrellina capilosa

Bryozoa
Cryptostomata

Like Trepostomata, Cryptostomata is a completely extinct order of bryozoans confined entirely to Palaeozoic seas. As distinct from the massive colonies (zoaria) of Trepostomata, cryptostomates formed intricate, reticulate colonies which sometimes look like petrified lace. The calcareous zoaria of Cryptostomata were formed of main branches, which either had variously constructed cross-connections or were connected to each other by undulating curves. The mouths of the individual chambers (zooecia) are usually localized on the branches but are occasionally on the cross-connections also, in two or more rows. Cryptostomatous bryozoans generally occur in calcareous sediments, and it is not uncommon to find their lacy zoaria in the company of corals and other reef-forming animals. The entire group appeared in the early Ordovician, spread over the whole globe during the Devonian and Carboniferous and died out in the Permian period.

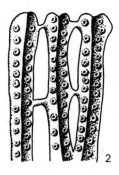

One of the most typical and also most prolific genera of Cryptostomata is *Fenestrellina*, whose reticulate zoarium was shaped like a fan or a funnel. The two rows of zooecia are situated on the main branches only, and the cross-connections (dissepiments) are unmarked. This genus, which lived from the Devonian to the Permian, is known all over the world. *Fenestrellina capilosa* (1) is a distinctive fossil of early Devonian limestomes of the Koněprusy reef in Bohemia.

Details of the structure of the colony, i.e. part of the main branches, with two rows of zooecia, can be seen in the enlarged illustration of part of a zoarium of the allied species *Fenestella antiqua* (2) from English Silurian strata.

The reticulate zoaria of *Reteporina prisca* (3) are differently constructed, the main branches, carrying the zooecia, being connected to each other in undulating curves, without any cross-connections (4). This species is typical of shallow-water limestones in German Devonian strata. Related species, differing only in minor details, abound in other parts of Europe, North America and Asia in lower Devonian to lower Carboniferous strata.

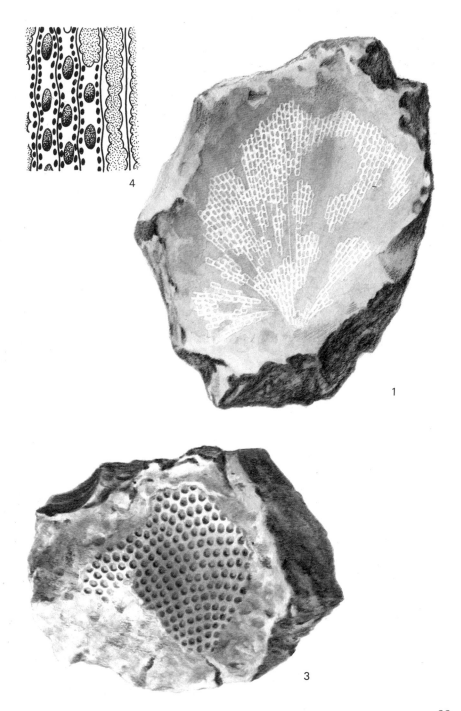

Lingulobolus feistmanteli

Brachiopoda
Inarticulata

The history of brachiopods dates back from the present to the early Cambrian. They are exclusively marine animals, and their body is covered with a shell formed of two valves of unequal size. The two parts of the shell are connected — in the case of the class Inarticulata by muscles only and in Articulata by locking devices of varying complexity, with teeth along the edge of one valve and corresponding grooves along the edge of the other. Brachiopods live at various depths, anchored to a base by a flexible stalk (pedicle). Some species are attached by the whole under side of the shell; some live within burrows, while others lie free on the sea bed. They live on microscopic life, which they drive towards their mouth by their whirling ciliated arms (hence the name Brachiopoda = arm-feet). The great era of the brachiopods was the Palaeozoic, and their remains make excellent index fossils for determining the age of Palaeozoic strata. In the Mesozoic era they lost much of their importance, and only a few species have survived to the present day. Members of the class Inarticulata are easily identified by their simply constructed shells.

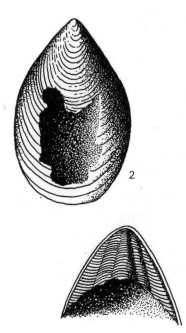

One of the largest members of the class Inarticulata is *Lingulobolus*, which is typical of Cambrian and lower Ordovician strata in Asia, Europe and North America. The valves, measuring up to 5 cm in length, have the form of a rounded triangle with a smooth surface marked only by fine growth lines. *Lingulobolus feistmanteli* (1) is a good example.

The genus *Lingulella*, which was distributed all over the world during the Cambrian and Ordovician periods and resembled the still extant genus *Lingula*, had typical lingulid, i.e. tongue-shaped, shells (2 — pedicle valve; 3 — detail of inner surface of apical part of valve). *Lingulella insons* (4) is a common fossil in early Ordovician slate strata in Bohemia. Similar species occur in Scandinavia, for example, and as far eastwards as China. The individual species of this genus are hard to differentiate without a knowledge of their internal organization, however, and so their names *(Lingula, Lingulella)* are used more or less collectively.

The tiny, rounded shells of the genus *Obolus* are likewise to be found all over the world, in Cambro-Ordovician sediments. The surface of the shells of *Obolus complexus* (5) is characteristically decorated with concentric growth lines combined with fine radial ribs. The inner mould of the shells shows a dinstinct central groove and two, paired impressions of muscles near the apex.

Jivinella incola

Brachiopoda
Articulata

Articulate brachiopods are very different from inarticulate types. They have calcareous shells formed of two valves of unequal size, the ventral (pedicle) valve generally being the larger and the dorsal (brachial) valve the smaller. At the apex of the posterior part of the shell there is a locking device, which, together with muscles, allowed the valves to be opened and closed. The whirling arms were not always free as in Inarticulata, but were supported as a rule by a variously constructed brachial apparatus. Articulates are marine animals, and their long geological history goes back from the present to the beginning of the Palaeozoic era. The superfamily Orthacea is one of the oldest groups. From the early Cambrian to the Devonian period, articulates abounded in seas all over the world. Since they lived in the most diverse environments, their remains are found in every type of sediment — clay, sand and limestone.

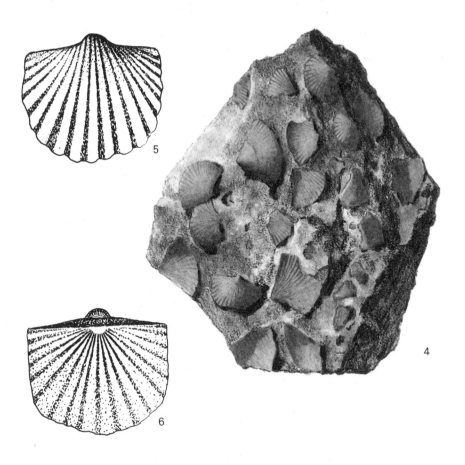

Billingsella is one of the oldest and most characteristic orthid brachiopod genera. It comes from middle Cambrian strata in North America, and its representatives have flat, semicircular, radially ribbed shells with small apices and a long, locking margin (hinge line). The diagram shows the ventral (1) and dorsal (2) valve of the typical species *Billingsella coloradoensis*.

The genus *Jivinella* from lower Ordovician strata in Bohemia is closely related to *Billingsella*. The ribbing on its valves is coarser and the apex is more pronounced, as seen in *Jivinella incola* (3).

Nisusia kuthani (4) was a shallow water type of articulate brachiopod. Its small shells, with their fine radial ribbing, are common in middle Cambrian sand and conglomerate horizons in Bohemia. Similar species are known from other parts of Europe and from Siberia.

The typical genus *Orthis*, after which the whole superfamily is named, is of early Ordovician origin and occurs chiefly in northern Europe and the USA. Its shells have conspicuous radial ribs and large teeth on the locking margin. *Orthis callactis* (5 — ventral valve; 6 — dorsal valve with straight locking margin) was already described in 1828 by Dalman, who found it in early Ordovician limestone sediments in the Baltic region.

Sieberella sieberi

Brachiopoda
Articulata

The superfamily Pentameracea comprises completely extinct brachiopods which inhabited shallow littoral waters mainly during the Silurian and Devonian periods. Most of them have large, massive, highly convex valves, with inturned apices and a short locking margin (hinge line). On the surface of the valves there are large numbers of pronounced radial ribs. These brachiopods abound in limestone and tuffaceous-limestone sediments all over the globe, and they are sometimes so numerous that their shells are an important component of the rocks. Examples of this are the Bohemian and English Silurian brachiopod strata containing the species *Gypidula caduca,* etc. and known as conchidial limestones.

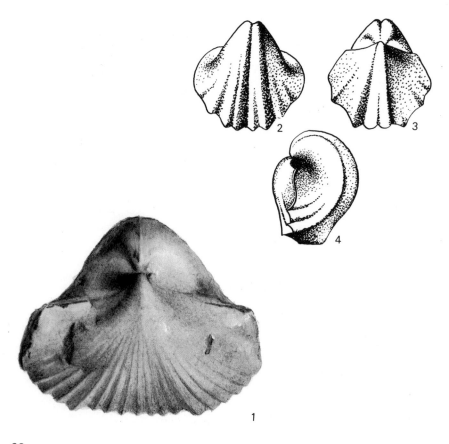

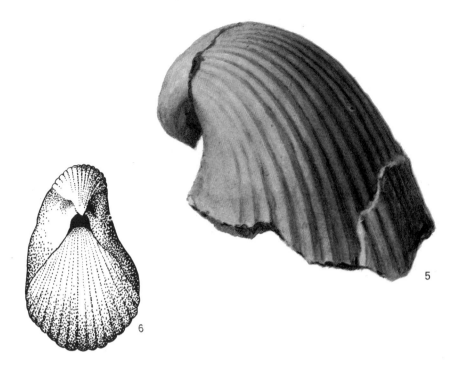

5

6

Siberella, one of the best known genera of pentamerid brachiopods, is distributed in Devonian limestones throughout the world. The species *Sieberella sieberi* (1), which has sharply ribbed shells up to 5 cm long, is plentiful in early Devonian organic débris limestones which settled along the margins of the coral reef near Koněprusy in Bohemia.

Other common genera are *Gypidula* and the very similar *Ivdelinia*, which has smaller, but characteristic shells. *Ivdelinia procerula*, from Bohemian lower Devonian strata, has a bulging, helmet-shaped ventral (pedicle) valve which is much larger (2) than the relatively flat dorsal (brachial) valve (3). Fig. 4 gives a side view of the two valves. On both of them the ribbing is stronger in the middle and fades towards the margins. The above genera are well represented in Silurian and Devonian limestones all over the world.

The largest pentamerid brachiopods were the numerous species of the equally widely distributed genus *Conchidium*, whose massive shells are up to 10 cm long. The markedly convex ventral valve (5), which has an inturned apex, is particularly large. The dorsal valve is also extremely vaulted, but is much smaller. This can be seen from Fig. 6, which shows the dorsal valve, the apex of the ventral valve and the opening for the stalk (pedicle) by which the animal is anchored. All the known species of the genus *Conchidium* come from Silurian and lower Devonian strata only. Often they are an important component of coral associations in Silurian tuffaceous limestones, e.g. in Bohemia, England and the USA.

Leptaena depressa

Brachiopoda
Articulata

The articulate brachiopods of the suborder Strophomenidina are a remarkable and exclusively Palaeozoic group. Although they lived only from the Ordovician to the Triassic period, they managed to spread to all the great oceans. Their characteristic and easily identified shells have an almost semicircular or semi-elliptical form and the sides are often drawn out like wings; the apex is small and there is a long and usually straight locking margin (hinge-line). The surface of the valves is adorned with radial or concentric ribs and grooves (usually with both elements combined). As distinct from most brachiopods, whose shells have two convex valves, Strophomenidina have a convex ventral (pedicle) valve and a concave dorsal (brachial) valve, so that both valves curve in the same direction. These brachiopods generally lie loose on the muddy sea bottom in places with a gentle current, but the other extreme is also known, i.e. species attached by the whole under surface of their ventral valve to foreign bodies or stones in high energy environments.

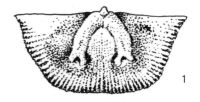

1

Aegiromena aquila is one of the geologically oldest species. It abounds in middle Ordovician shales over Europe, north Africa, South America and Asia. The internal moulds of valves illustrated here also show the inner structure, i.e. the smaller mould of the dorsal valve, the curved supporting ridges for the whirling arms (brachia) (1), the larger mould of the ventral valve and the kidney-shaped (reniform) muscle impressions below the apex (2).

The genus *Leptaena*, whose massive, only slightly curved, semicircular shells turn down almost at right angles at the edges, is typical of the Silurian and Devonian periods. These brachiopods lived in clean, flowing water and their remains are found, therefore, chiefly in limestone sediments. *Leptaena depressa* (3) is a common fossil of European Silurian strata; similar species are to be found practically all over the globe.

Cymostrophia stefani (4) is a geologically more recent, specialized species whose large, long-winged shells are quite plentiful in places and are typical of early Devonian reef limestones near Koněprusy in Bohemia. The characteristic lattice pattern on the surface of the shells is produced by a combination of radial and concentric ribbing. Similar species have been found in southeastern Europe and north Africa, but they are still being studied.

Chonetes tardus

Brachiopoda
Articulata

Many specialized and curiously shaped types are to be found among the articulate brachiopods. This applies especially to members of the families Chonetidae and Productidae, which date back to the early Palaeozoic but reached their peak in the Carboniferous and Permian periods, i.e. the late Palaeozoic, when they inhabited all the great oceans. The surface of their often bizarrely formed shells is frequently covered with sharp ribs and spines of different lengths. The size of the shells also varies, and some of them measure up to 15 cm across. The individual species have a characteristic structure and wide geographical distribution, so they are excellent index fossils, especially for dating Carboniferous and Permian sediments. Some species probably lived attached to the sea bed, while others lay free, anchored only by their spines in the soft ooze. Chonetidae already appear in upper Ordovician and lower Silurian strata, Productidae in lower Devonian strata. At the end of the Palaeozoic, in the Permian period, they both suddenly disappeared, and in Mesozoic seas there are no signs of either of them.

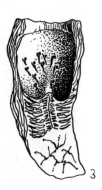

Chonetes tardus (1), from Silurian limestones in central Bohemia, is a typical chonetid brachiopod. The surface of its semicircular, winged valves has sharp, radial ribbing, and on their straight locking margin (hinge line) there are four long, hollow spines.

The long spines on the ventral (pedicle) valve of the large productid *Cancrinella altissima* (2), which comes from upper Carboniferous layers in USA, kept the animal anchored within the shallow sea bed. The spines were brittle, and most of those found were broken.

Members of the productid genus *Richthofenia*, known from Permian strata in Europe, southeastern Asia, and north Africa had interestingly constructed shells. Since they lived in places with strong currents, they have a massive, conical ventral valve. Fig. 3 shows a section of the valve of the typical species *Richthofenia lawrenciana* from

Permian limestones in Pakistan. The valve was attached to the substrate not only by the tip of the cone, but also by root-like spines. The dorsal (brachial) valve is flat and is no more than a kind of small lid; the appearance of its inner surface can be seen in Fig. 4.

Eospirifer togatus

Brachiopoda
Articulata

The order Spiriferida comprises extinct brachiopods with highly specialized and intricate internal structures supporting their 'mobile' arms. These calcareous brachial structures are shaped like two symmetrical, three-dimensional spirals with the apices turned inwards or towards the margins of the valves. The valves themselves are likewise typical and are either smooth or, more often, have sharp radial ribs. In the centre of the larger ventral (pedicle) valve there is usually a trough-like depression (sulcus) matching a wide, rounded or sharp-edged fold of varying height in the middle of the dorsal (brachial) valve. The countless species of spiriferid brachiopods lived from the Ordovician period of the Palaeozoic era to the Jurassic period in the Mesozoic. Their shells are known in every part of the world, and often their remains have accumulated to form an important component of various rocks, especially limestones. These brachiopods lived mainly in shallow littoral regions with flowing, well-aerated water. Their frequently huge shells were attached to a base by a strong, flexible stalk (pedicle).

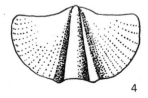

4

Eospirifer togatus (1), which has large, broadly oval, smooth or finely grooved valves with a single wide, rounded fold down the centre, is one of the best known representatives of the spiriferid brachiopods. It is a characteristic fossil of early Devonian limestones in central Europe and in Asia. Similar species lived in North America during the Silurian and Devonian periods.

Brachiopods of the genus *Hysterolites,* e.g. *Hysterolites nerei* (2, 3) are a different type, but are likewise index fossils; they occur in lower Devonian strata especially in Germany and are distributed all over the world. Their triangular shells are widest along the hinge line, where the valves lock together (2), and as well as a high, sharp central fold they have several other radial ribs (3).

The members of the genus *Cyrtia* have a low, normally shaped dorsal valve (4), but the ventral valve is like a tall pyramid cut vertically in half (5). They are confined to Silurian and Devonian formations, but occur all over the globe. *Cyrtia exporrecta,* the species illustrated here, is a characteristic fossil of European Silurian limestones.

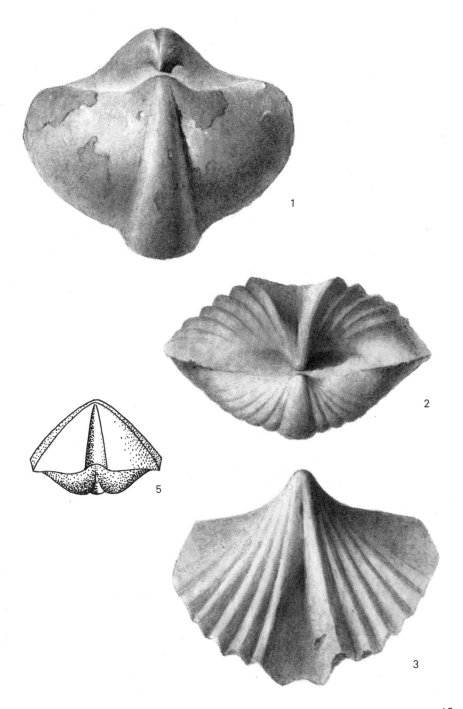

'Atrypa' verneuiliana

Brachiopoda
Articulata

Few of the articulate brachiopods frequently found in masses in every type of Palaeozoic sediment have a really distinctive form. The majority have the usual biconvex valves with pronounced apices and a short, usually curved locking margin (hinge line), the surface of the valves being smooth or only faintly sculptured. These brachiopods are generally hard to study, because of the similarity of their outward appearance. It requires serial sections and thin sections to reveal the tremendous diversity of the inner structure of the shells and hence the full profusion of the often biologically very different species of these inconspicuous groups. These seemingly simply constructed brachiopods include both the members of the completely extinct suborder Atrypidina (known from the Ordovician to the Devonian period) and more modernly organized brachiopods, like the members of the suborder Terebratulidina, which have survived in the seas from Devonian times to the present day.

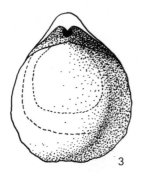

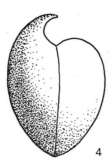

The collective genus 'Atrypa', with a series of species known from Silurian and Devonian layers all over the world, is a typical and well-known member of the first of the two suborders named above. Its biconvex valves measure up to 5 cm and the ventral (pedicle) valve is always noticeably less convex than the dorsal (brachial) valve. The surface of both valves is adorned with rounded ribs crossed by concentric lamellae. The whole forms a distinctive lattice, with elongate scales where the ribs and lamellae cross. 'Atrypa' verneuiliana (1) is a characteristic fossil of European lower Devonian limestones. The Silurian species 'Atrypa' renitens (2) has smooth valves.

With its massive, almost spherical shells the size of a lemon, Stringocephalus burtini is one of the most distinctive and stratigraphically most valuable terebratulid brachiopods (suborder Terebratulidina). The ventral valve is much bigger than the dorsal valve and has a large, beak-like apex with the orifice (foramen) for the stalk (pedicle) below it (3). The beak-like apex is also striking when seen from the side (4).

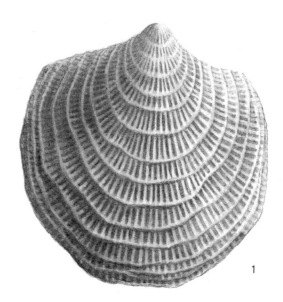

1

Stringocephalus lived in the shallow parts of the European middle Devonian seas, attached to the firm calcareous beds by a flexible stalk (pedicle). The rocks are often so full of their shells that they are known as 'stringocephalus' limestones.

2

Acanthochitona calliozena

Mollusca
Polyplacophora

These bilaterally symmetrical molluscs known as coat of mail shells or chitons, with an elongate case formed of eight overlapping shell-plates (valves) on the dorsal surface, are encountered in practically all marine areas. Few people know that they are an ancient and very conservative class of molluscs and that their fossilized remains can be found in upper Cambrian marine sediments. In the more than 500 million years of their existence their mode of life has changed little. They crawl slowly over the sea bed in the shallow littoral zone (although deep sea species are also known), feeding on algae. Fossil species presumably lived in the same manner. The order Palaeoloricata includes more primitive types which lived mainly during the Palaeozoic era and died out in the Cretaceous period. Extant species belong to the evolutionarily higher order of the Neoloricata, whose first members appeared in the Carboniferous period. Fossil remains of polyplacophorans are somewhat rare and are usually found as fragments, though morphologically clearly identifiable. The shell-plates are held together only by muscles, so that when the animal dies the shell usually breaks up into the separate segments.

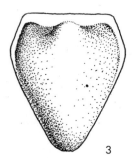

Acanthochitona calliozena (1), a typical representative of recent polyplacophorans, exemplifies the appearance and organization of the armour of extinct species. The differently shaped plates at the two ends — the rounded head (cephalic) plate and the bluntly triangular tail (apical) plate — and the roof-like intermediate plates are characteristic. This genus, whose known history goes back to Tertiary sediments, has persisted to the present day in the majority of seas.

In Silurian shallow-water limestones and tuffaceous sediments in central Europe, Scandinavia, Great Britain and the USA, isolated plates of the genus *Chelodes*, up to several centimetres long, are occasionally found. The tail plates of *Chelodes bohemicus* (2) in particular, which taper to a long spike, are typical and are easily identified also by the thick growth lines on their surface.

The genus *Eochelodes* is geologically older than the preceding species, from which it differs also in other respects. The plates are small and smooth and are shaped like a blunt-angled pentagon (3). They occur in slates, indicating that this species tended to frequent the ooze on the quieter and deeper parts of the sea bed. This genus is likewise known from central Europe and Scandinavia and was quite recently discovered in the USA and Australia.

Retipilina knighti

Mollusca
Monoplacophora

Monoplacophorans are a very primitive group of marine molluscs with a bilaterally symmetrical body covered on the dorsal surface with a hood- or spoon-like shell. It had always been presumed that the entire group was completely extinct and that its occurrence was confined to the early Palaeozoic, from the Cambrian to the Devonian period. In 1955, however, the Galathea expedition from Denmark fished up a living monoplacophoran from the depths of the Pacific Ocean. The astounding discovery of this famous 'living fossil' — which was described as *Neopilina galatheae* — not only provided information on the organization of the soft parts of these molluscs' bodies, but also made possible their exact classification in the animal system as a whole. Monoplacophorans had previously been virtually classified as a kind of snail (Gastropoda), but they were found to be a specialized, evolutionarily isolated group allied rather to the worms. Further finds of living *Neopilina* confirmed that these living fossils' way of life had hardly changed over a period of almost 500 million years. They crawl slowly over the sea bed and live on organic débris. Fossil types chiefly frequented shallow water, whereas recent *Neopilina* are known only from the depths of the ocean.

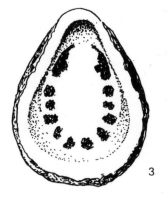

In Silurian tuffaceous limestones in Bohemia we occasionally find massive, spoon-like shells conspicuous for their marked, rounded, concentric ridges and for the reticular pattern on their surface. They belong to the distinctive species *Retipilina knighti* (1), whose members lived only on the firm tuffaceous-sandy bed of shallow parts of the Silurian seas.

Another species with a similar shell and habits is *Tryblidium reticulatum*, which is characteristic of middle Silurian limestones on the Swedish island of Gotland. It is the typical species of the whole order. *Tryblidium* shells have sharp, concentric ribs with prominent lamellae (2). Fig. 3 shows an inner view of the shell with the prominent muscle impressions.

Upper Silurian limestones in Europe and north Africa contain different species of monoplacophorans which Barrande named *Drahomíra* (a girl's name). Only fragments of the smooth, thin-walled shell have been preserved, and little is known beyond the spoon-shaped inner core, on which, however, seven pairs of symmetrical inner muscle impressions are clearly visible (4). *Drahomíra* travelled slowly over the muddy bed of quiet creeks, where strong currents were absent.

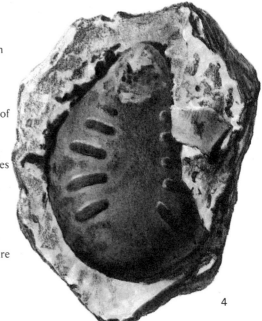

Cardiola docens Mollusca
Bivalvia

Early Palaeozoic sediments contain large quantities of very characteristic bivalve mollusc shells (Bivalvia) whose locking device was either so fine that it was destroyed, or had only poorly developed teeth. The systematic classification of these simply constructed, undifferentiated shells is difficult, but they seem to belong to types from which some evolutionary lines of recent molluscs sprang. They were therefore grouped in a somewhat artificial order — Palaeoconcha, now called Cryptodonta — and only detailed study and time will be able to tell which of the members of the order are genuinely related and which bear only a morphological resemblance to each other. Cryptodonta (= Palaeoconcha) already appeared in the upper Cambrian, abounded in the Silurian and Devonian periods and suddenly declined at the end of the Palaeozoic era. As a point of taxonomic interest, many of these molluscs which come from Silurian and Devonian formations in central Bohemia have Czech, and not Latin, generic names (e. g. *Panenka, Královna, Vlasta, Maminka, Tenká*, etc.), which were given them by the famous French palaeontologist Barrande.

One of the best-known genera of the subclass Cryptodonta (= Palaeoconcha) distributed in Silurian sediments in Europe, Asia, north Africa and Australia is *Cardiola* (1 — *Cardiola docens*), whose concentric growth lines are crossed by radial ribs, forming a rough lattice ornamentation on the surface of the shell. The wide geographical distribution of some species is related to their mode of life, since they seem to have attached themselves by special filaments (byssal threads) to seaweed or to the shells of cephalopods and to have been carried considerable distances by the ocean currents.

Another typical genus is *Panenka*, many species of which occur in Silurian and Devonian rocks in Europe and North America. These bivalves also have thin-walled, oval shells with pronounced radial ribs. *Panenka expansa* (2) is a characteristic fossil of early Devonian formations in central and southern Europe.

The shells of *Praecardium* (3) are generally smaller than those of *Panenka*, but have characteristic conspicuous radial ribs, which are rounded or angular in cross section. Their remains are found mainly in limestones in most parts of the world. *Praecardium* seems to have crawled on the sea bed or burrowed in the soft sediment.

53

Ctenodonta bohemica

Mollusca
Bivalvia

The subclass Palaeotaxodonta likewise comprises ancient and primitive types of molluscs with a specifically characteristic — taxodont — locking device (hence the name of the order). The device consists of a row of stud-like teeth usually arranged perpendicularly or obliquely to the locking margin of the shell. The members of this order inhabited early Palaeozoic seas in large numbers and their remains are used as index fossils, especially for the Ordovician. Some species eventually migrated to brackish and fresh water, and their remains are found in Permian and Carboniferous freshwater lake sediments, for instance, together with other freshwater fauna and particularly with the typical flora that eventually produced anthracite. Taxodont bivalves are still to be found in the seas today, but they are the last residue of what was once a huge and prolific order.

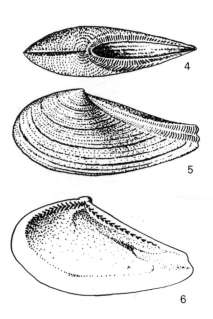

Middle Ordovician sediments in North America, southeastern Asia and especially central and southern Europe are characterized by the typical oval shells — or more frequently the inner moulds — of small palaeotaxodont bivalves of the genus *Ctenodonta*. The moulds in particular show the highly typical locking device and anterior and posterior muscle impressions in the now dissolved shell. *Ctenodonta bohemica* (1 — from the side, 2 — from above) is just one example of the many species of this genus. All ctenodonts lived and crawled on the sandy-argillaceous beds of cold Ordovician seas.

Phestia, which has an extremely elongate shell and is known chiefly from brackish and freshwater Permian-Carboniferous sediments in Europe, Asia and North America, is a younger type of palaeotaxodont bivalve. *Phestia attenulata* (3) inhabited the muddy bed of Carboniferous lakes and to some extent the less salty creeks of seas now occupied by the European mainland.

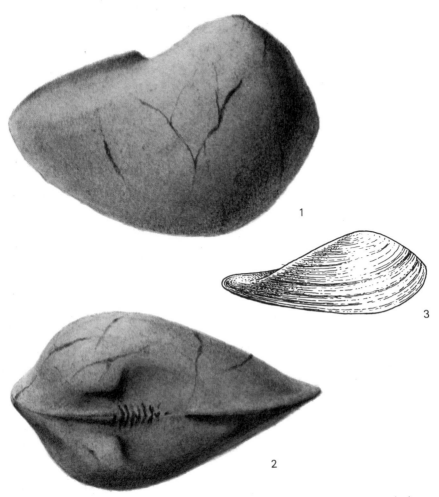

One of the most resistant types is the nut clam *Nuculana,* which is known all over the world and has managed to survive, in practically the same form, from the Devonian period to the present day. The slender, club-shaped shells of the recent species *Nuculana pernula* (4, 5, 6) show the basic organization of the shell and the structure of the locking device (hinge teeth) and also indicate the way in which the nut clams' ancestors undoubtedly lived all those millions of years ago. It is no exaggeration to describe these creatures as living fossils.

Aviculopecten multiplicans

Mollusca
Bivalvia

Among the infinite number of species of bivalves which inhabited Palaeozoic seas, there is another artificially determined group, known from the Ordovician to the present. This is the 'order' Dysodonta, whose members have a usually asymmetrical shell with the apex near the front; the locking part (hinge) is produced to short or long wings and the teeth have been reduced to studs or ridges, as in oysters *(Ostrea)*, for example. Although they did not attain maximum incidence until the Mesozoic and Cainozoic eras, their attractive remains belong to the typical fossils of Palaeozoic sediments, and they often have considerable biostratigraphic significance. The majority are sessile and live attached by strong byssal threads to a fixed or floating object, or one of the valves of their shell may be cemented to a rock on the sea bed or the side of a cliff. Conversely, some species are able to swim by a kind of jet propulsion achieved by rapid opening and closing of the valves, as in the case of the scallops (Pectinidae).

The species *Aviculopecten multiplicans* (1) lived in flowing, well oxygenated water on the margin of the lower Devonian coral reef near Koněprusy in Bohemia. Its relatively large valves, which have a small apex and indistinct wings, abound in the limestone where the currents washed them into the stiller parts of the original reef.

Leiopteria mucro (2), from upper Silurian, fine, detrital bituminous limestones in Bohemia, is another remarkable species. It is the representative of a genus known from the Silurian to the Permian period and widely distributed in Europe and North America. *Leiopteria* valves are small and triangular, with the apex well to the front; the wings are asymmetrical, the anterior one being small and the posterior one long and conspicuous. The shape and occurrence of these molluscs suggest that they probably led an epiplanktonic existence, i.e. attached by their byssal threads to floating objects like seaweed.

Among the representatives of the dysodont bivalves which are useful for biostratigraphy, the members of the worldwide genus *Posidonia*, such as the typical *Posidonia becheri* (3), are of primary importance. Their large, obliquely oval, thin-walled valves with poorly developed wings and pronounced concentric ribs, are particularly common in lower Carboniferous shales named 'posidonia shales' after them.

Boiotremus fortis

Mollusca
Gastropoda

The innumerable species of gastropods, generally known as snails and slugs, occur in salt and fresh water and also on dry land. Their body is protected by a single, unpaired calcareous shell, which is usually twisted in a kind of spiral, although it may be shaped like a hood or cornet, or reduced altogether. Gastropods have a well differentiated head with eyes and tentacles, and a muscular foot adapted for crawling or swimming. Aquatic (and especially marine) species are typical bottom-dwellers. Their shells are generally strong and are preserved well in the rocks. Gastropods are a very old group and had already appeared in the Cambrian period, at the beginning of the Palaeozoic era. The most successful in Palaeozoic seas were the members of the primitive prosobranch (= 'front-gilled') order Archaeogastropoda, which evolutionally are also the oldest. It is worth noting that a number of still extant species of archaeogastropods — living fossils with a family tree going back over 500 million years — still live in the seas of today.

Among the most primitive and geologically oldest species are the members of the genus *Helcionella*, which had a simple, hood-like shell whose surface was decorated with concentric ribs. *Helcionella subrugosa* (1) is known from early Cambrian strata in the USA.

In another group within the order Archaeogastropoda — the bellerophontids — the typical genus *Bellerophon*, represented by the species *Bellerophon vasulites* from German middle Devonian strata, is characterized by an almost spherical, tightly coiled shell, on which the last coil almost completely covers the older coils (2, 4). The opening of the shell has a shallow notch for the anal siphon (3). From the Silurian to the Devonian period, this was a common genus all over the world.

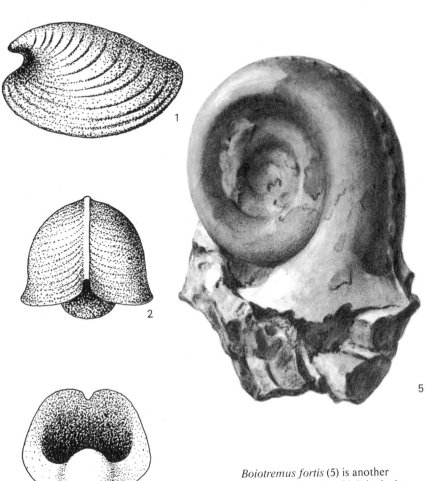

Boiotremus fortis (5) is another distinctive bellerophontid. It had a large, flat to discoid shell, however, with a trumpet-like orifice. The notch for the anal siphon closed periodically during growth and, instead of one, there are therefore several openings.

The genus *Tremanotus,* known from Ordovician and Silurian strata in various countries of Europe, Australia and North America, closely resembled *Boiotremus.* The only difference is that in the latter the openings for the anal siphon ran the full length of all the coils, while in *Tremanotus* they occurred only in the last, adult coil (Fig. 6 shows *Tremanotus civis*).

Platyceras gregarium

Mollusca
Gastropoda

The platyceratacean gastropods are an interesting and abundant group of exclusively Palaeozoic archaeogastropods, with large, massive shells up to 15 cm high. The older (innermost) coils of the shells are small, but their size quickly increases and the last (youngest) spiral is very wide. Loosely coiled and conical shells are also known, however. Platyceratacean gastropods lived mostly on the muddy bottom of shallow, quieter parts of the sea, but they did not shun places with strong currents, such as the margins of coral reefs. They do not seem to have been very active. Some species lived symbiotically attached to the top of the cups of large crinoids (sea lilies), and in Europe and particularly in the USA there have been a number of finds of Silurian and Devonian crinoids with their attendant platycerataceans, which probably lived on the excreta of these echinoderms. Free platyceratacean gastropod shells, however, are known from every part of the world; they occur in various types of Ordovician to Permian sediments, though mostly in limestone.

The representatives of the Silurian and Devonian subgenus *Orthonychia*, which occur in various parts of the world, are allied to the genus *Platyceras*. Today they are held to comprise platyceratacean gastropods with partly or completely uncoiled, conical or hooded shells. *Platyceras (Orthonychia) anguis* (2), whose shell is coiled in a loose spiral, is one of the more interesting species.

Another species, *Platyceras (Orthonychia) elegans* (3), is remarkable for several reasons. It has an unusually tall, conical shell up to 15 cm high, with undulating walls. It lived attached to the top of the cups of large, planktonic *Scyphocrinites* crinoids and was carried by its hosts to every part of the world.

Platyceras gregarium (1) belongs to the typical genus, which had a tightly coiled, quickly widening shell. Its large, hard shells are a common fossil constituent of early Devonian reef limestones near Koněprusy in Bohemia. Similar species are known from other parts of Europe and from Asia and North America.

61

Oriostoma eximium

Mollusca
Gastropoda

The primitive and rather quaint archaeogastropods were not the only members of the class Gastropoda which lived in early Palaeozoic seas. The majority of gastropod species were probably almost the same as regards appearance, shell form, body organization and habits as the marine snails of today. The significant features for systematics are the shape of the shell, the way in which it is coiled and surface adornments such as spines, outgrowths, ridges or ribs. In a number of species, the shell opening could be closed with a massive lid (operculum), as it is today, and both fossil opercula and whole, closed shells are known. These ancient gastropods are most abundant in Ordovician, Silurian and Devonian sediments. In the later Palaeozoic their numbers decline, and they are replaced by more modern species. However, isolated species still survive today, in a slightly modified form, as living fossils.

The members of the genus *Oriostoma*, which had wide, conical, spiral shells, were important inhabitants of the shallow parts of Silurian and Devonian seas in Europe and North America. In some of them the surface of the shell was variously adorned, but in every case it had a massive operculum. The typical species *Oriostoma eximium* (1 — shell, 2 — operculum interior, 3 — operculum exterior) is a common fossil of European Silurian tuffaceous limestones.

The small gastropods of the genus *Tubina* inhabited shallow, well-oxygenated water in the vicinity of reefs. Their shell is coiled in a regular flat spiral, while the apex is decorated with rows of straight or curved spines and

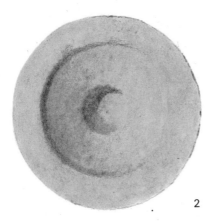

2

3

spikes of varying sizes. The oldest (initial) coils are small, while the last (adult) is large and loose. In this way the shell offered better resistance to the currents. *Tubina armata* (4, 5) is one of the many species characteristic of early Devonian formations in Europe, North America and Asia.

Among the evolutionally more advanced types of gastropods we find some with tall, 'towering' shells and noticeably long apertures. The surface of the shells was smooth or marked only with fine growth grooves. The typical species *Loxonema sinuosum* (6) is characteristic of middle Silurian strata in England, while similar species occur in somewhat younger Silurian and early Devonian sediments in central and eastern Europe and southeastern Asia.

Endoceras sp.

Mollusca
Cephalopoda

Cephalopods are exclusively marine molluscs. A part of the foot has been transformed to eight, ten or more tentacles arranged on their head (hence their name, which means head-feet) in a ring round the mouth. The rest of the foot has been converted to a kind of funnel, the hyponome, on the under side of the body. The funnel is directed forwards, and water is expelled through it with such force that the animal shoots backwards, rather like a jet plane in reverse. Of all the evolutionary branches of the molluscs, the cephalopods have attained the highest degree of perfection as regards both the efficiency of certain organs (e.g. the eyes) and the organization of the nervous system and hence perception. The majority are active swimmers and have inhabited all the seas and oceans since the late Cambrian, i.e. for at least 500 million years. Most recent species have a large, calcareous, inner shell, such as the familiar cuttle-bone, whereas fossil types had external shells whose form varied from a slender cone to a spiral. The various successive groups of cephalopods played an important role in Palaeozoic and Mesozoic seas in particular, and today their remains make excellent index fossils.

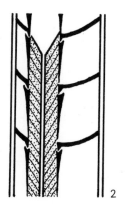

Among the oldest Palaeozoic cephalopods are the members of the specialized subclass Endoceratoidea, whose many genera are known from Ordovician limestone layers in northern Europe and Asia and in North America. Some species attained a considerable size, with conical shells up to 9 m long. The longitudinal sections (1, 2) and cross section (3) of a specimen of *Endoceras* sp. from North American Ordovician limestones shows the basic structure of the shell, which had simple septa dividing off the air chambers (camerae) and an eccentric longitudinal axis, the siphonal tube (siphuncle), through which ran the blood vessels and neuromuscular tracts. The wide siphonal tube was partly filled with calcareous, conical capsules fitting inside each other; these are known as endocones and are typical of the entire subclass.

Endoceratoids were prolific in the warm Ordovician seas of the Baltic-Scandinavian province, where their shells formed whole cephalopod limestone positions, whereas in the cold Mediterranean palaeoprovince (now occupied by central and southern Europe and north Africa) they were rare. They did stray there occasionally, however, as seen from the shell and distinctive siphuncle of the species *Endoceras novator* (4) found in early Ordovician strata in Bohemia and France.

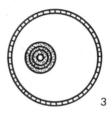

'Orthoceras' arion

Mollusca
Cephalopoda

Nautiloidea are the only subclass of Palaeozoic cephalods still represented in recent seas. The familiar living fossil known as the Pearly Nautilus lives in the Indian Ocean and is the last member of a subclass which attained the peak of its development in the Ordovician and Silurian periods. In the late Palaeozoic, Nautiloidea suddenly declined, and at the beginning of the Mesozoic all types with a straight shell ('orthoconic') suddenly vanished, so from the Mesozoic to the present only species with a spiral shell are encountered. In the modern systematic classification of Palaeozoic nautiloids, attention is paid primarily to the structure of the shells and their inner components as revealed by normal and ground sections. Since these are difficult and expensive methods for the ordinary collector, he keeps to the old, rough form of classification by the shape of the shell.

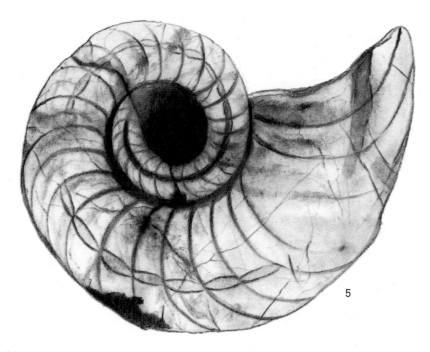

5

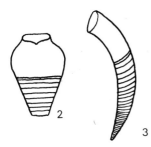

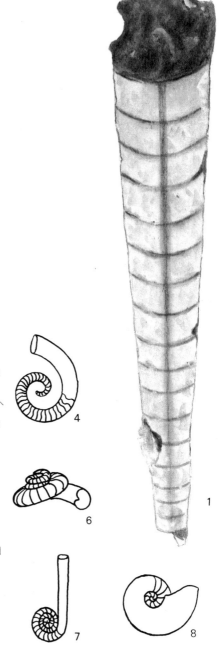

The section of the shell of the species *'Orthoceras' arion* (1) shows the structure of the straight, conical (orthoconic) shell, which has a central siphuncle and simple, curved septa between the air chambers (camerae). The animal lived in the youngest chamber, which opened via the aperture.

Robust, straight, widely conical and relatively short cells with a fairly narrow aperture are described as brevicones (2). The Silurian genus *Rizosceras* is an example.

Variously curved shells, usually with an excentric siphon, are known as cyrtoconic (3). This is a very common type of shell, especially among Silurian nautiloids.

If the shell forms a free, open spiral in a single plane, (e.g. in the Devonian genus *Ptenoceras*), it is termed gyrocone (4).

Shells forming a tight spiral are nautilicone (serpenticone), as in *'Barrandeoceras' bohemicum* (5).

Torticonic shells form three-dimensional spirals of varying heights, as in the Silurian genus *Peismoceras* (6).

Lituiticone shells first of all formed a tight spiral, but the last coil straightened out, so that the shell as a whole looked rather like a bishop's crook (7). The Silurian genus *Ophioceras* is an example.

The modern nautilus shell is a convolute type; the younger body spirals partly overlap and protect the older spirals with the air chambers (8).

Bathmoceras praeposternum

Mollusca
Cephalopoda

Early investigators attached systematic importance to the shape of the various types of nautiloid cephalopod shells and therefore distinguished genera like *'Orthoceras', 'Cyrtoceras'* and *'Trochoceras',* etc. Today we employ these only as collective, cumulative names and put them in inverted commas, since they do not express evolutionary relationships, but only convergence, i.e. similar morphological adaptation of different evolutionary lines as a result of the effect of external and internal factors such as the same mode of life. For instance, the oldest and most primitive members of the order Ellesmerocerida, which appeared in the late Cambrian and died out during the Ordovician, and the more modernly built Michelinoceratida, which were common during the whole of the Palaeozoic and survived into the Mesozoic, both had straight shells. The rocks are often crammed full of Silurian michelinoceratid shells collected and deposited by the current, and, as orthocerid limestones, they were often exploited as material for decorative purposes.

3

The European *Bathmoceras praeposternum* (1) is one of the most interesting ellesmerocerids. It had a large dwelling chamber, narrow air chambers (camerae) packed tightly together and almost straight, flat septa drawn conically forwards only in the centre, beside the siphonal tube (siphuncle). This species is a distinctive fossil of the cold European Ordovician sea of the Mediterranean palaeoprovince.

One of the most remarkable michelinoceratids is the Silurian *Parakionoceras*, which had long, slender shells whose surface was decorated with regular, fine, longitudinal ribs. *Parakionoceras originale* (2) abounded in European upper Silurian seas. Similar genera are known from limestones of the same age in North America, Asia and Australia.

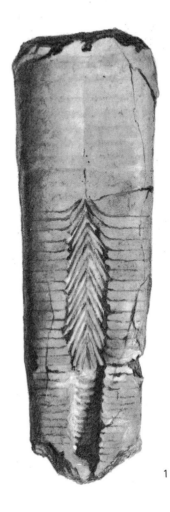

1

2

The diagram (3) shows the biological position and basic organization of the body of orthoconic nautiloids. The white spaces are the air chambers, separated by simple concave septa. This part of the shell was actually a complex hydrostatic organ allowing the animal to rise or sink or to float freely in the water.

Corbuloceras corbulatum

Mollusca
Cephalopoda

Although the commonest Palaeozoic nautiloids were those with an orthoconic shell, in Silurian sediments in particular are found different species with shells of the various other types. The Silurian was in general a time of great development of these ancient cephalopods, and, consequently, although thousands of species are known from all over the world, modern research on these animals is far from being complete. It is remarkable that, especially in the Silurian, the appearance of differently constructed shells was accompanied by a wide range of sculptures on their surface, e.g. longitudinal and transverse ribbing (costae and annulations), regular rounded ridges (lirae) and rows of various rings and protuberances (nodes). Nautiloids found in Silurian limestones also yield information on the colour patterns with which the shells of some species were covered and which were preserved on the fossil remains in the form of undulating or zig-zag lines, longitudinal and transverse bands or regularly arranged spots.

Corbuloceras corbulatum (1), from European upper Silurian limestones, is an example of a species with a cyrtoconic shell and a distinctively sculptured surface. The shell is about 12 cm long and not more than 5 cm wide. Its surface is covered with prominent, longitudinal ribs

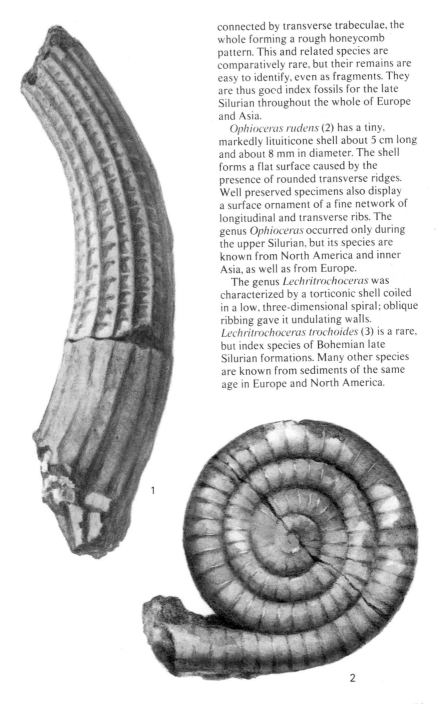

connected by transverse trabeculae, the whole forming a rough honeycomb pattern. This and related species are comparatively rare, but their remains are easy to identify, even as fragments. They are thus good index fossils for the late Silurian throughout the whole of Europe and Asia.

Ophioceras rudens (2) has a tiny, markedly lituiticone shell about 5 cm long and about 8 mm in diameter. The shell forms a flat surface caused by the presence of rounded transverse ridges. Well preserved specimens also display a surface ornament of a fine network of longitudinal and transverse ribs. The genus *Ophioceras* occurred only during the upper Silurian, but its species are known from North America and inner Asia, as well as from Europe.

The genus *Lechritrochoceras* was characterized by a torticonic shell coiled in a low, three-dimensional spiral; oblique ribbing gave it undulating walls. *Lechritrochoceras trochoides* (3) is a rare, but index species of Bohemian late Silurian formations. Many other species are known from sediments of the same age in Europe and North America.

Octamerella calistomoides

Mollusca
Cephalopoda

In the great majority of Palaeozoic nautiloid cephalopods, the mouth (aperture) of the dwelling chamber, and hence of the whole shell, was completely open; in section it was circular or oval. Along with these species with a simple mouth, are found, especially in Silurian and Devonian seas, specialized cephalopods in which the mouth of the shell was partially closed in a variety of ways. In every case, however, a larger upper, and often regularly lobate, opening, connected by a narrow space with a smaller, simple opening in the lower part of the narrowed mouth, can be distinguished. The animal thrust its tentacles through the upper, lobate opening, while the lower orifice was the opening for the hyponome. These curiously formed nautiloids had been noticed by early research workers, who placed them in an artificially determined group, Microphaga, on the assumption that they were non-predatory species living on microscopic plankton. This classification, together with the theory of their eating habits, no longer applies. These specialized species lived on the same food as other nautiloids with a narrow-mouthed shell, and the only difference is that they were better protected from their enemies.

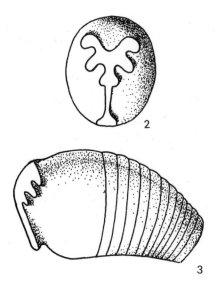

The species *Octamerella calistomoides* (1) had a short, breviconic shell with a very complicated mouth. The upper orifice had eight regular lobes for the individual tentacles, while the lower opening (for the hyponome) projected forwards into a kind of hood.

Hexameroceras panderi (2 — anterior view; 3 — side view) had a similarly formed, narrow mouth, but with only six lobes for the tentacles. The vertical space connecting the opening for the tentacles with the opening for the hyponome was long and straight. Like the rest of the genus *Hexameroceras*, this is a relatively rare species, but the palaeogeographic distribution of the genus as a whole is fairly wide and it is known from inner Asia and China, as well as from the Silurian of Europe and North America.

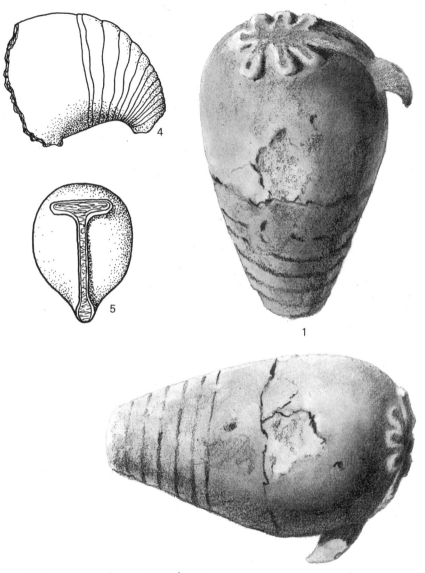

Bolloceras, which has a cyrtoconic or gyroconic shell up to 25 cm long, with a trumpet-like mouth (4), is another remarkable genus. The mouth of the shell is narrowed by two side lobes; these divide the oval upper opening for the tentacles, which is connected by only a narrow space with the hood-like opening for the hyponome (5). *Bolloceras rex* and several other species are characteristic fossils of Devonian limestones in Europe and North America.

'Barrandeoceras' bohemicum

Mollusca
Cephalopoda

The best organized species were those with a nautiliconic or convolute shell coiled in a solid spiral, with the younger (wider) coils partly (convolute) or completely (involute) covering the older (smaller) ones. This type of shell is already found among Silurian nautiloid species, and, as the structurally and evolutionarily most progressive, it persisted through the Mesozoic and Cainozoic eras to the present-day genus *Nautilus*, which is the last representative of the subclass Nautiloidea. None of the less progressive types of shell was successful in geologically more recent periods, since their bearers all died out in the Palaeozoic era. In fact, not even species with nautiloconic shells can be regarded as direct ancestors of the present-day *Nautilus*, as was once thought to be the case; they are only different, and often completely extinct, morphologically convergent types in the phylogenetic development of nautiloid cephalopods during the past 500 million years.

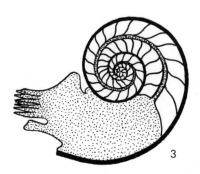

The extant *Nautilus pompilius* (1) from the Indian Ocean is the last relic of these once widespread 'rulers of the early Palaeozoic seas'. It lives at a depth of several hundred metres and we therefore usually find only its empty shells (measuring up to 25 cm) washed up on to the shore by the breakers.

The Palaeozoic cumulative genus *Barrandeoceras*, which used to be included in the genus *Nautilus*, has a similar form and a similarly constructed shell. Among the many species known from Silurian formations in Europe, North America and north Africa, the most important is *'Barrandeoceras' bohemicum* (2), whose shells, which measure up to 30 cm, are occasionally found in Silurian limestones in Bohemia, France and Morocco.

The diagrammatic section of the body of a nautiloconic cephalopod in the biological (i.e. living) position shows how

2

the soft body lies in the dwelling chamber; the siphuncle (not in line with the middle of the body) and the air chambers (camerae), protected against damage by the younger coils of the spiral shell (3), are unshaded.

1

Anarcestes plebejus

Mollusca
Cephalopoda

Ammonites (Ammonoidea) are more intricately organized cephalopods which probably broke away from the evolutionarily simpler nautiloids some time during the late Silurian. The oldest ammonites, which appeared at the beginning of the Devonian period, still had a straight or curved calcareous shell divided by inner septa hardly any more complex than those of nautiloids. The first ammonoids, with a typical spiral shell and increasingly complex contact lines between the air chambers ('goniatites'), appeared during the early Devonian. During the late Palaeozoic the development of the ammonoids proceeded apace, and in the Mesozoic era they ruled all the seas and oceans and every type of marine environment from deep to shallow waters and from cold subarctic to tropical waters. At the end of the Mesozoic, a break occurred in the evolution of the ammonoids, and their innumerable species suddenly disappeared. The shells of Palaeozoic and Mesozoic ammonoids, like those of nautiloids, are often an important component of rocks and can form very thick cephalopod limestone horizons. In addition, ammonoids are of primary importance for biostratigraphy (strata dating). Most of them were active swimmers, and individual species spread rapidly over extensive areas. This, together with the explosive development of the whole subclass, in which more and more new species evolved in a geologically short space of time, makes ammonoids excellent index fossils.

3

The most primitive ammonoids are of the suborder Anarcestina, which have a simple, rounded shell coiled in a spiral. They are known only from Devonian strata, but appear in large numbers throughout the whole of Europe and Asia and in north Africa. *Anarcestes plebejus* (1), a small species known from early Devonian limestones in different parts of Europe, is a typical representative of Anarcestina.

The similar genus *Mimagoniatites*, likewise known from early and middle Devonian formations in Europe and north Africa, was characterized by a slim, discoid shell. To judge from the shape of its shell, *Mimagoniatites fecundus* (2) was a good, fast swimmer and probably frequented the deeper parts of the sea. It

is a common fossil in European middle Devonian beds.

The distinctive species *Pinacites jugleri* (3) from European middle Devonian strata illustrates the characteristic suture lines termed goniatic.

Gonioclymenia speciosa

Mollusca
Cephalopoda

In the late Devonian, a specialized group of ammonoids classified in a separate order, Clymeniida, broke away from the primitive anarcestids. The shell of *Clymenia* was coiled in a tight spiral and had simple, undulating septal suture lines between the air chambers (camerae). The striking difference between the Clymeniida and all other groups, however, was the position of the siphonal tube (siphuncle), which was situated along the dorsal (inner) margin of the shell and not along the inner side of the ventral (outer) wall. The entire order appeared at the end of the middle Devonian, developed dramatically in the late Devonian and suddenly disappeared everywhere at the end of the Devonian period. The various species are thus of immense value for detailed stratigraphic division of upper (late) Devonian strata. Their shells are found chiefly in limestones and calcareous sediments, are easily identified and are sometimes so abundant that they form horizons of clymeniid limestones. The majority of species have been found in Europe, inner Asia and north Africa; on other continents they are rarer.

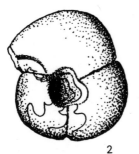

2

3

Gonioclymenia speciosa (1) represents the phylogenetically most important types of *Clymenia*, from which further, more modern groups evolved. It has a narrow, discoid shell up to 15 cm in diameter, with slightly curved, rounded ribs on the surface. The septal contact lines (sutures) are markedly and sharply undulating. This species is characteristic of German upper Devonian strata; related species are known from the whole of Europe and from north Africa.

The small members of another index genus, *Wocklumeria*, from European and north African upper Devonian limestones, are of a more modern construction. A side view of the German species *Wocklumeria sphaeroides* (2) shows the typical shape of the shell, where the coils are so close together that the younger ones almost completely cover the older ones (involute). The surface of the wide, rounded shell is constricted regularly three times.

A longitudinal section of the shell (3) shows its internal structure. The simple air chamber septa, which describe a sharp curve on the dorsal wall in places where the siphuncle was situated, are typical.

Ceratites semipartitus

Mollusca
Cephalopoda

At the end of the Palaeozoic era, in the Permian period, the old, declining and dying goniatites were replaced in seas all over the world by a new order of ammonoids — Ceratitida — which heralded a new period of explosive development in the history of the ammonoid cephalopods. The shells of ceratites were coiled in the normal, tight spiral, and the surface was smooth, simply ribbed or decorated with rows of protuberances (nodes or tubercles). What differentiated them from other ammonoids was the course of the septal contact lines (sutures) between the air chambers, which is so characteristic in the majority of species that they are known as ceratitic sutures. The lines are undulating as in goniatites, but only the anteriorly directed saddles are simple; the posteriorly directed lobes are broken up by small, sharp arcs. Ceratites inhabited all the seas everywhere in the Permian period and during the beginning of the Mesozoic, i.e. the Triassic period, when they dominated, but when they were in turn superseded by phylogenetically higher types of ammonites. They were excellent swimmers and evidently preferred the open sea. Today their remains are found mostly in sediments deposited in very deep regions.

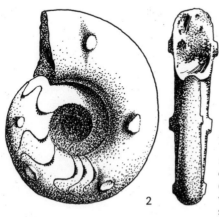

European — and especially German Triassic — limestones contain numerous species of the typical genus *Ceratites*. One of the commonest, *Ceratites semipartitus* (1), had a shell up to 20 cm in diameter. In weathered individuals, the typical ceratite sutures are clearly discernible. The genus itself is something of an exception, as it is confined to Europe and, unlike other ceratites, seems to have frequented the shallow parts of the sea.

Ceratites belonging to the genus *Tirolites* have an almost worldwide incidence. Their rounded, relatively wide shells measuring up to 10 cm in diameter are surmounted by regular and sometimes strikingly prominent protuberances (tubercles). They are known from early Triassic sediments in the Alps, the Carpathians, the Balkans and the Caucasus in Europe and also from he Himalayas, southeastern China and the USA. *Tirolites idrianus* (2) is an index fossil for lower Triassic strata in the south of Europe.

Xenodiscus plicatus, which had a flat shell with rounded edges (3 — side view; 4 — front view) and typical ceratite sutures (5), is one of the oldest (late Palaeozoic) ceratites. The species is characteristic of upper Permian sediments in India and Indonesia.

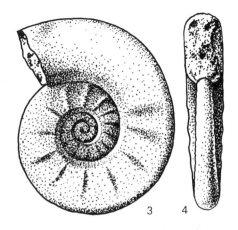

Proplacenticeras orbignyanum

Mollusca
Cephalopoda

The true ammonites were the last and the most highly developed stage in the evolution of the ammonoid cephalopods. Whereas the most characteristic animals of the early Palaeozoic era were the trilobites, the Mesozoic seas and oceans were dominated by these cephalopods. During the Jurassic and Cretaceous periods in particular, many related or only morphologically similar branches of ammonites developed, giving rise to thousands of species. At the end of the Cretaceous they suddenly declined, however, and not one of them survived into the Cainozoic era. The systematics of the true ammonites are very complicated and this survey, therefore, keeps to the feature which characterizes them as a group, i.e. the structure of the sutures (the contact of the septa between the air chambers with the shell wall). In true ammonites the suture was intricate, its individual lobes and saddles broken up into often exceedingly complicated subsidiary curves. These complex 'ammonite' sutures were functional and made the shell extremely strong. The shells were predominantly of the familiar spiral type — smooth, flat and discoid, with sharp edges, in species whose members were probably good, fast swimmers, or wide and rounded, without a keel, and with thick ribs or protuberances (nodes or tubercles) on the surface. The latter type offered considerable water resistance and their owners therefore swam slowly, or let themselves be carried by the currents, like plankton. Both types spread rapidly to all seas everywhere, and today their shells are ideal index fossils.

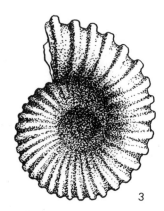

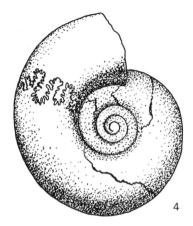

The members of the genus *Proplacenticeras* were fast, agile species; the shell, which measured about 10 cm in diameter, had a smooth surface. *Proplacenticeras orbignyanum* (1) and other related species abounded in upper Cretaceous seas in Europe, the USA and inner Asia.

Mantelliceras mantelli was a representative of the second type. Its wide, rounded shell was densely covered with prominent, tranverse ribs (2 — front view; 3 — side view). This widespread genus is known from lower Cretaceous sediments all over the world.

The classic ammonite suture still occurred occasionally in the worldwide Triassic ammonites of the genus *Gymnites*. *Gymnites incultus* (4), found in Triassic limestones in the Alps, Balkans and Himalayas, is an example.

Collignoniceras woollgari

Mollusca
Cephalopoda

Whereas in Triassic and Jurassic ammonites the shape of the shell was relatively stabilized in the functionally most satisfactory tight spiral form, in the Cretaceous period degenerative features became increasingly common. They included gigantism (shells up to 2.5 m in diameter) and fancy spirals. Some shells were half-uncoiled, others hooked, others again had deformed or irregular spirals. Some species with straight shells actually reverted to the phylogenetically primitive Palaeozoic 'orthoconic' type. These morphological changes seem also to have denoted a change in the mode of life of these curious species. Judging by the shape of the shells the owners can hardly have been capable of fast, active movement; they probably let themselves drift with the currents and were thus dependent upon the supply of microscopic food or they may have been benthic. These features are all indicative of a general decline of the ammonites, which culminated in loss of their dominant role among the Cretaceous invertebrate marine fauna and in their complete disappearance at the end of the Mesozoic era.

Collignoniceras woollgari (1), an upper Cretaceous European ammonite, had a wide, massive shell over 1 m in diameter, with thick ribs and protuberances (tubercles) on its surface. The genus as a whole occurs worldwide.

Scaphites had a remarkable shell. The older coils form a tight spiral, but the shell then straightens out, and the

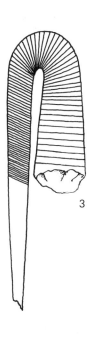

youngest part curves back like a hook. *Scaphites* abound mainly in the uppermost (youngest) Cretaceous layers, but some species already lived in early Cretaceous seas, e.g. *Macroscaphites yvani* (2), known from Europe and north Africa.

The species *Anahamulina subcylindrica* (3), from European and North American Cretaceous formations, had a curiously shaped shell. The older part is straight, while the younger part, with the dwelling chamber, is recurved like a hook and spread out like the mouth of a trumpet.

Some types of middle Cretaceous ammonites even had shells of a gastropod type in appearance, such as *Turrilites costatus* (4), a member of a genus known from Europe, the USA, Africa and India. In fact, this type of ammonite was originally considered to be a gastropod. The small, tapering, spiral, sharp-tipped shell was about 10 cm high with an elongate mouth, and the surface was decorated with regular transverse ribs.

Nowakia cancellata and *Styliolina fissurella*   Mollusca (?)
Tentaculitida

A group whose systematic position is uncertain

Already in the nineteenth century, small, narrowly conical or needle-like shells, only a few millimetres long, were found in places covering complete shale and limestone bedding planes. They were named 'Tentaculites' after their needle-like form, but nobody bothered to study them in detail. It was only comparatively recently that 'tentaculitids' were found to be of considerable stratigraphic significance and interest in them was restimulated. As there are no impressions of their soft bodies, it is not known where they belong in the animal system. Once they were classified among the still extant sea slugs (Opisthobranchiata), i.e. among the molluscs, but lately they have come to be regarded as a group allied more to the annelid worms. They are extinct marine animals with fine calcareous shells and their occurrence is confined to the Silurian and the Devonian periods. They were capable of free movement, and some species seem to have lived near the sea bed in shallow water round the coasts of the continents, while others swam on the surface of the great oceans. They soon spread all over the world, and today tentaculitids can be used for the exact dating of Silurian, and especially Devonian, sediments everywhere.

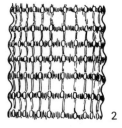 2

 3

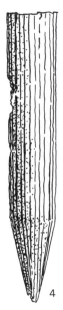

The commonest tentaculitid species in Devonian shales and limestones belong to the two cosmopolitan genera *Nowakia* and *Styliolina* (1), whose representatives often appear together. The shells of *Nowakia* species are larger and conical, with transverse rings and fine, dense, longitudinal ribbing on their surface (2, 3). The shells of *Styliolina* are narrower and smaller (about 3 mm long) and have a smooth surface.

The genus *Striatostyliolina* is related to the genus *Styliolina*, but differs from it in respect of the longitudinal ribbing on the surface of its needle-like shells and the slightly swollen embryonic cavity in the tip of the shell. *Striatostyliolina peneaui* (4) is plentiful in middle Devonian formations in Europe, north Africa and inner Asia.

Maxilites maximus

Mollusca (?)
Hyolitha
A group whose systematic position is uncertain

Hyolitha is another interesting class of marine animals with bilaterally symmetrical, conical shells. Its members lived only during the Palaeozoic era, from the Cambrian to the Permian, and attained their maximum distribution right at the outset, i.e. in the Cambrian period. We must assume, therefore, that they already had a long evolution behind them, stretching back into the Proterozoic era. They had larger shells than the tiny tentaculitids, the average length being 3 cm and the maximum 15 cm; in section they had the form of a round-cornered triangle. The apex of the hyolith shell was divided inside by septa, and the mouth had an operculum and a pair of adoral appendages. On the inner side of the opercula are variously shaped locking devices and impressions of the muscles which kept the operculum in place over the shell mouth. The long, recurved adoral processes led from the shell on either side, between the operculum and the mouth, and probably acted as stabilizers. The habits of hyoliths have not yet been fully clarified, but they seem to have crawled over the muddy bed in deep, quiet areas of the sea.

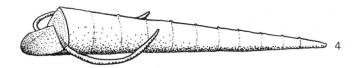

Maxilites maximus was one of the largest hyoliths, and its shell (1), which has a simple operculum (2), measures about 10 cm. It is a comparatively rare species and is found in middle Cambrian slates in southern and southeastern Europe.

The unusually lucky find of a complete specimen of *Pauxilites solitarius* (3) in Bohemian middle Ordovician slates shows the flat under side of the shell, the operculum, complete with teeth and processes, and the recurved adoral appendages. From this find, a complete hyolith of the genus *Pauxilites* in life position (4) was reconstructed. This is a fairly common species in European Ordovician formations, and allied species have been found in north Africa and southeastern Asia.

Eurypterus fischeri

Arthropoda
Eurypterida

The Palaeozoic era, especially its early part, is sometimes called the age of invertebrates. The name is very apt, because Palaeozoic sediments contain the remains of diverse and often weird invertebrates which lived during this era only and disappeared when it came to an end. One of these specifically Palaeozoic groups is the subclass Eurypterida, the collective name for the largest arthropods (segmented animals) in the Earth's history. Some of them attained huge dimensions and measured over 3 m. Since they were all predacious, they were undoubtedly a menace to other animals. Eurypterida had an elongate, segmented body encased in chitinous armour, with a large, half oval cephalothorax (combined head and thorax), opisthoma (abdomen) and a variously formed, flat or tapering caudal segment (telson). On the upper side (dorsal) of the cephalothorax there was a pair of large, compound eyes, of the same type as in insects, and behind them small, simple eyes (ocelli). On the under side of the cephalothorax there were six pairs of appendages, transformed either into oral appendages (sometimes with a pair of powerful pincers) or a locomotor apparatus for swimming. Phylogenetically older species lived in the sea, apparently on or near the bottom in shallow creeks. Younger species successively migrated to brackish or fresh water on the various continents.

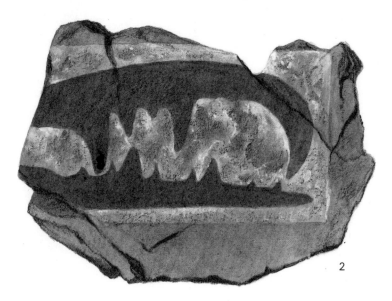

2

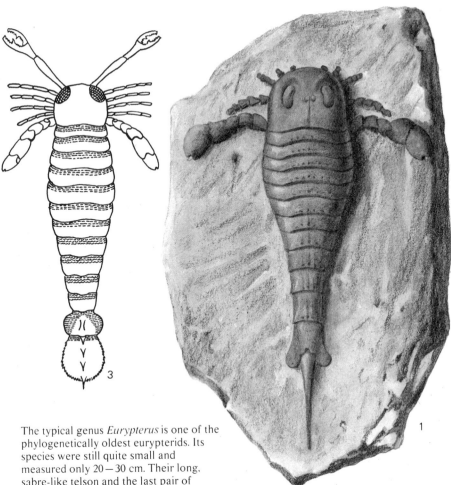

The typical genus *Eurypterus* is one of the phylogenetically oldest eurypterids. Its species were still quite small and measured only 20—30 cm. Their long, sabre-like telson and the last pair of appendages on their cephalothorax, which had been converted to powerful rowing organs, indicate that they were good swimmers. They lived primarily in the Ordovician and Silurian seas of Europe, North America and Asia, but range up to the Carboniferous. The index species *Eurypterus fischeri* (1) is common in Baltic-Scandinavian Silurian formations.

The members of the Silurian-Devonian genus *Pterygotus* measured up to 3 m in length. Their shell terminated in a wide, flat telson, and their cephalothorax was equipped with a pair of powerful, spiny pincers. On the surface of the shell was a characteristic design of small semicircles, so even fragments are easily identified. Isolated pieces of shell of the species *Pterygotus bohemicus* (2), including pincers up to 20 cm long, are found from time to time in upper Silurian limestones in Bohemia.

Remains of the similar species *Pterygotus buffaloensis* (3) are found in sediments of the same age in the USA. Further species of the same genus occur in Spitzbergen and in Australia.

Aristozoe memoranda

Arthropoda
Crustacea

Specialized crustaceans belonging to the subclass Phyllocarida, which display signs of close relationship to the extant marine order Leptostraca, form another group of arthropods which abounded in Palaeozoic seas. In this respect, leptostracans can, therefore, also be regarded as living fossils. As in most crustaceans, the body of phyllocarids was protected by a segmented chitinous exoskeleton. The most frequently found fossilized part is the carapace, the large, oval part of the shell covering the thoracic segments and the first segments of the abdomen. The carapace is usually split into two identical, convex halves, rather like an ordinary, half-filled purse. Anteriorly they are joined by the movable rostral plate covering the head. The posterior part of the carapace is linked to 1—7 abdominal segments terminating in a long, spiked telson, which sometimes has two lateral spikes. Phyllocarids were evidently good swimmers. They lived in every part of the sea, not excluding the shallow surf zone in the vicinity of coral reefs. Their fossils date from the early Cambrian and are known all over the globe. They attained maximum distribution during the Silurian and Devonian periods and then, in the later Palaeozoic, the whole group slowly but steadily started to decline and all that is left to remind us of them today are representatives of the order Leptostraca.

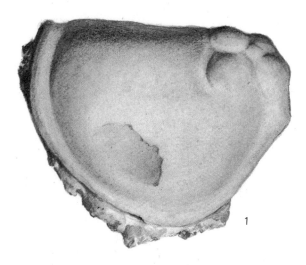

In Silurian, and especially Devonian, European limestones we sometimes find the oval carapaces, 8 cm long, of the genus *Aristozoe*, with a straight dorsal margin and a distinct border. The anterior part of the shell is covered with a variable number of characteristic rounded protuberances (nodes).

Aristozoe memoranda (1) is a small species with a carapace about 4 cm long. It is a characteristic fossil of lower Devonian limestones bordering the Koněprusy reef in Bohemia.

Ceratiocaris is a very widely distributed phyllocarid genus, and its members are known from practically all over the world and for the whole of the Palaeozoic era, i.e. from the Cambrian to the Devonian period. Here and there in Bohemian Silurian limestones are found large numbers of characteristically formed telsons about 10 cm long, belonging to the species *Ceratiocaris bohemicus* (2), which look like spines with longitudinal ribs and rows of fine thorns on their surface.

The reconstruction of the related species *Ceratiocaris papilio* (3) from Silurian formations in Scotland shows the general organization of the shell of ceratiocaridids, together with the limbs.

2

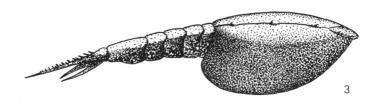

3

Harperopsis bohemica

Arthropoda
Crustacea

Ostracoda are one of the scientifically most important crustaceans classes, although they are not very popular with collectors. They have lived on our planet since the early Cambrian period, i.e. the beginning of the Palaeozoic, and are still here today. The majority are minute, and their body is completely encased in an oval, bivalve and usually thoroughly calcified shell. The surface of the valves of Palaeozoic ostracods is often decorated with ribs or outgrowths, while geologically younger and extant forms are generally smooth. Ostracods swim actively and live in vast numbers in the sea, in fresh water and even in the wet soil of tropical forests. Because of their frequent mass occurrence in all types of sedimentary rocks, their wide geographical distribution, but small time span (and hence stratigraphic range) of individual species, they are extremely useful in micropalaeontological research. They require highly specialized study, however, and we include here only a few informative examples of the many thousands of species already known.

3 ♀

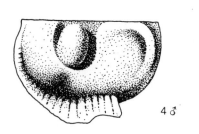

4 ♂

The shells of the genus *Harperopsis*, are only a few millimetres long and are characterized by four vertical, rounded ribs on the surface of the valves. They ar typical of Ordovician strata in Europe and North America. In some cases they literally cover bedding planes in slates, as in the case of *Harperopsis bohemica* (1), a distinctive middle Ordovician fossil.

Other typical ostracod species can be found in Silurian and Devonian fine-grained limestones and marls. The valves have oval swellings on their surface, or bristles with variously shaped spikes and outgrowths. These striking

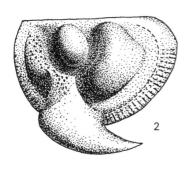

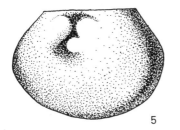

types include the members of the typical genus of the *Beyrichia*, which is represented in Europe by the species *Beyrichia latispinosa* (2), for example, and in North America by the related species *Beyrichia moodeyi* (3 — a female; 4 — a male).

The members of the purely European genus *Parapyxion* have completely smooth or only slightly sculptured, oval valves with a straight upper hinge line. They are characteristic fossils of middle and upper Ordovician sediments. The typical species *Parapyxion subovatum* (5) is known chiefly from Scandinavia.

Marella splendens

Arthropoda
Trilobitoidea

In the nineteenth century, some curiously shaped fossils were found in Bohemian middle Ordovician quartzites and siliceous sands. They had the form of a central irregular square or triangle, from which four, long, recurved, flat processes always protruded. For a long time researchers were unable to agree as to where the mysterious remains should be classified, as they did not resemble any of the then known groups of organisms. Barrande finally described them as *Furca bohemica* (because of their fork-like appearance), on the assumption that they were the caudal appendages of some kind of crustacean. It was not until 1909 that the mystery was eventually solved, when C.D. Walcott discovered and described the rich and beautifully preserved fauna of the middle Cambrian Burgess shales of British Columbia. Among this fauna he discovered numerous complete specimens of small arthropods with a similarly constructed body to trilobites, but with a forked head plate (cephalon) like the one described above. He described them as *Marella splendens* and they were found to be members of a large group of primitive aquatic arthropods known as Trilobitoidea, remote relatives of the trilobites. Trilobitoids lived during the Palaeozoic era only and their various lines evolved in different ways.

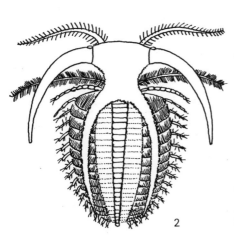

2

The illustrations (1, 2) show the appearance and body organization of *Marella splendens,* the typical representative of these remarkable arthropods. They seem to have frequented nothing but the muddy parts of Cambrian seas, where there were no strong currents. Members of this genus have so far been found only in Canada (in the Burgess shales).

Furca bohemica is a marellid-like arthropod, and the unique remains found in Bohemian Ordovician strata are not caudal appendages, but the head plates of a large species of these arthropods (3). Unlike the typical genus, however, the remains of *Furca bohemica* occur solely in sandy sedimentary rocks of littoral origin. This species is typical of the Bohemian Ordovician, although similar problematical remains are known also from France and north Africa.

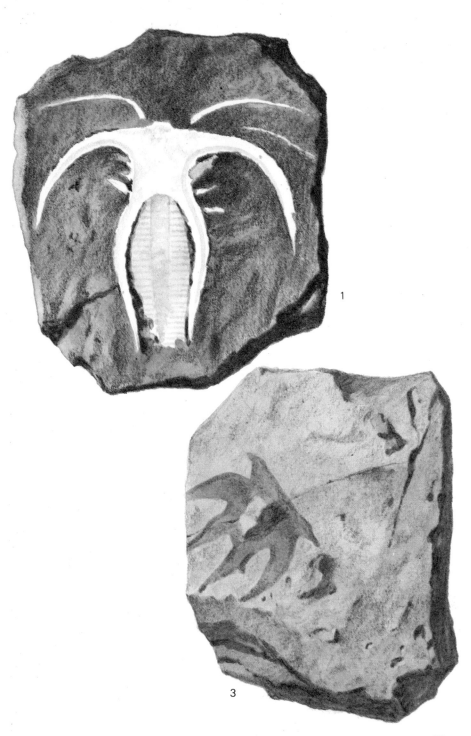

Paradoxides gracilis

Arthropoda
Trilobita

Trilobites (class Trilobita) are a phylogenetically compact group of exclusively marine Palaeozoic arthropods whose dorsal surface was covered with a hard, chitin-like exoskeleton. The carapace was usually flat, more or less oval and divided into three main parts — a head region *(cephalon)*, a body section *(thorax)* and a tail section *(pygidium)*. The middle, bulging part of the head plate, the glabella, was either smooth or had a few pairs of glabellar furrows or lobes. Trilobites usually did not exceed 5 cm, although species up to 75 cm long are also known.

Paradoxidids (superfamily Paradoxidacea) are one of the best known groups of trilobites; stratigraphically they are one of the most valuable and are very popular with collectors. They have a primitively constructed, elongate and relatively flat carapace with a large, semicircular cephalon, a slender thorax composed of 13—22 segments and a very small pygidium (micropygeus). The free cheeks of the cephalon, the thoracic segments (pleurae) and the marginal segments of the pygidium project to form recurved, sabre-like spines. The cephalon is surmounted by a characteristic large, pear-shaped glabella with two pairs of transverse glabellar furrows and moderately large convex eyes. The occurrence of paradoxidids is confined primarily to the middle Cambrian period, but they had a wide area of distribution and are known from Europe, the Atlantic coast of North America, Asia and Africa. They are consequently distinctive index fossils and, formerly, the middle Cambrian was called the paradoxidid stage after them. Their remains occur mainly in slates, shales and silts, i.e. sediments of the deeper, still parts of the sea.

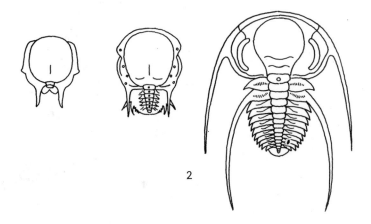

2

Almost complete specimens of the familiar typical species *Paradoxides gracilis* (1), with its long, slim carapace are known from Bohemian middle Cambrian strata.

The similar genus *Hydrocephalus* is known from other parts of Europe and from North America. It is differentiated from the typical genus by its wider, oval thorax and details of the head and pygidium. *Hydrocephalus carens* is famous as one of the first trilobites to have its ontogenetic development described, from the simple microscopic larva to the 30-cm adult individual (2).

Ellipsocephalus hoffi

Arthropoda
Trilobita

The trilobites of the superfamily Ellipsocephalacea, known from lower and (chiefly) middle Cambrian strata, form an exceedingly diverse group with worldwide distribution. They are characterized by a generally elongate carapace, a semicircular, relatively large cephalon, a large number (12—14) of thoracic segments (pleurae) and a tiny pygidium (micropygeus). The cephalon is conspicuous for the prominent glabella and eyes. These trilobites preferred the still, deeper parts of the sea, with gentle currents and a muddy bottom. Dozens of species are known; they come mainly from the northern hemisphere, although problematical remains have also been found in Australia and Argentina.

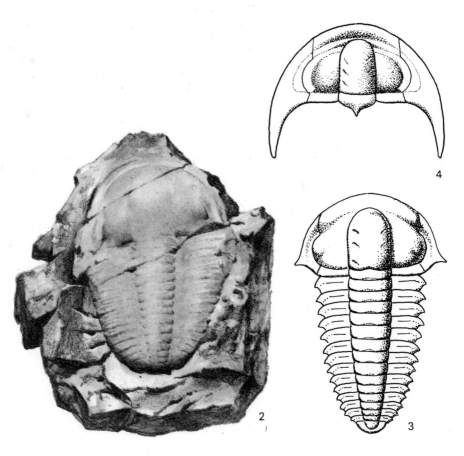

Ellipsocephalus hoffi (1) was first described by Schlotheim (1823) in Bohemia, and since then it has been collected all over the world. Not only is it the typical representative of the whole superfamily, but its carapace is often found complete and in large numbers. Most of them are exuviae, i.e. they represent shed carapaces which the currents carried away and re-deposited.

Germaropyge germari (2), which was formerly also included in the genus *Ellipsocephalus*, is a larger and more robust Bohemian middle Cambrian ellipsocephalacean. It differed from the slimmer, smaller *Ellipsocephalus hoffi* in its mode of life, as well as in appearance. *Germaropyge* rather frequented shallow water with a sandy bed, so its remains are common in sandstones; it is rare for it to be found in shales and slates. *Germaropyge germari* is an endemic species, that is to say, its occurrence is confined to the Bohemian middle Cambrian.

Ellipsostrenua gripi (3), a European early Cambrian species, had a slim, elongate carapace, a large cephalon and a tiny pygidium. The free cheeks on the lateral margins of the cephalon were each extended to form a small (genal) spine.

The likewise early Cambrian genus *Strenuella* (4), whose various species are known from Europe, north Africa and North America, had conspicuous free cheeks produced backwards to a curved (genal) spine, and a spiked occipital ring at the base of the glabella.

Phalagnostus nudus

Arthropoda
Trilobita

Most fossil enthusiasts tend to be attracted by the larger and more striking types of trilobites and to overlook apparently inconspicuous species like the members of the order Agnostida.

Agnostids were miniature, blind trilobites with a special body structure and mode of life. At a cursory glance, their shell seems to be composed of just two parts — a semicircular cephalon and a similarly formed pygidium of practically the same size. The two are sometimes so alike that collectors often do not know which end is which. The body (thorax) is very short and small and is composed of only two segments (very rarely of three).

All known agnostids are typical of the Cambrian and Ordovician periods. Their worldwide distribution is evidently associated with their way of life. The majority lived in clumps and beds of seaweed and were carried long distances by the ocean currents. Agnostid remains are found mostly in slates, shales and limestones deposited in moderately deep parts of the sea.

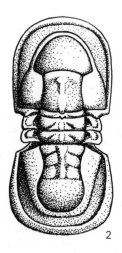

2

Phalagnostus nudus (1), with its simply constructed, semicircular cephalon, without a border or marginal thickening, is a typical agnostid trilobite. The pygidium has a short, indistinct axial spindle and a narrow border. This species has been found in middle Cambrian sediments in different parts of Europe and in Canada and Australia.

The largest known genus of the whole order is *Condylopyge* found in various European countries and in parts of North America. Individual specimens measure up to 1.5 cm and are easily distinguished by the triangular frontal lobe of the cephalon and the typical structure of the axis of the pygidium. This axis is composed of three rings and bulges towards the tip. *Condylopyge rex* (2) typifies the middle Cambrian.

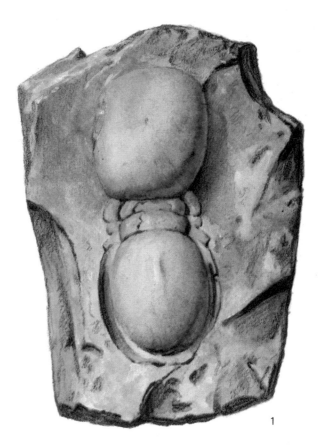

The genus *Peronopsis* (3), whose members are only 5—7 mm long, is known from European, Asian and North American formations. It is characterized by a glabella with a square frontal lobe and by the long, median axis of the pygidium.

Conocoryphe sulzeri

Arthropoda
Trilobita

The trilobite suborder Ptychopariina comprises hundreds of species distributed all over the world. They still have a primitive structure like the paradoxidids, for example, although the likeness ends there. Ptychopariina have regularly oval or ovoid, convex carapace with a semicircular or round-cornered, trapezoid cephalon, a large, wide thorax composed of 7—42 segments and mostly a small, wide and short pygidium. This morphology of the carapace documents the popular theory that phylogenetically old types of trilobites have a multi-segmented thorax and a small pygidium, while in modern types the thorax undergoes reduction and the pygidium grows steadily larger. Ptychopariid trilobites are known from lower Cambrian to upper Ordovician formations in every part of the world. Many species are excellent index fossils, especially for Cambrian dating.

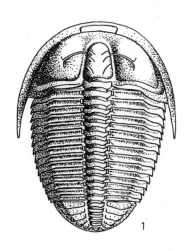

1

The members of the genus *Ptychoparia*, after which the whole suborder is named, have an ovoid carapace with sharp contours. The striking glabella has a rounded trapezoid form, 3—4 pairs of simple glabellar furrows and large eyes seated on prominent palpebral lobes. The occurrence of the various species of *Ptychoparia*, e.g. *Ptychoparia striata* (1), is confined to European middle Cambrian formations.

At first glance, the genus *Conocoryphe* looks very like the preceding genus, and collectors often mistake the one for the other. In the species *Conocoryphe sulzeri* (2), from the middle Cambrian in Bohemia, the differences are clearly discernible, however. *Conocoryphe sulzeri* has a conical glabella, with only three pairs of glabellar furrows, and — most important of all — it was blind. It probably burrowed in the fine sea bed, while *Ptychoparia striata* lived on the surface of the sea bed. The genus *Conocoryphe* is known only from European, North American and Asian Cambrian formations.

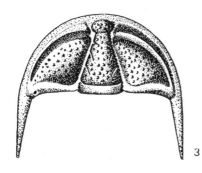

Ctenocephalus coronatus (3) is another related, but specialized species of ptychopariid trilobite. The margin of its semicircular cephalon is strikingly raised, forming a kind of peripheral barrier. Since this was also a blind species, it is quite likely that the purpose of the raised edge of the cephalon was to facilitate burrowing. This interesting species is a distinctive, but rare fossil in shale horizons in the middle Cambrian strata of Bohemia.

Sao hirsuta

Arthropoda
Trilobita

The superfamily Solenopleuracea likewise comprised more primitive types of trilobites characteristic of the earliest part of the Palaeozoic. They abounded chiefly in Cambrian seas and are known from all over the globe. They are distinguished by an oval or ovoid carapace with a semicircular or parabolic cephalon, a multi-segmented thorax (7—17 segments) and usually a small pygidium. On the cephalon there is a clearly circumscribed glabella with three (occassionally four) pairs of glabellar furrows. The remains of these trilobites are generally found in fine sediments, e.g. slates, silts or marly limestones, showing that the animals preferred to live on the muddy bed of the quieter parts of the sea. Some species were blind and these probably burrowed in the ooze.

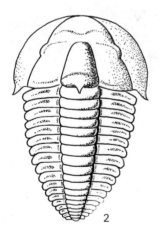

2

One of the best known solenopleurid trilobites is *Sao hirsuta* (1), whose carapace has sharp contours, a warty surface, a long thorax with 17 segments and a micropygeous pygidium with two segments. This species is remarkable chiefly for being the first trilobite to have the whole of its ontogenetic development described (by Barrande), from the smallest, almost microscopic larval stages (protaspid) to the adult individual (holaspid).

The tiny trilobite *Agraulos ceticephalus* (2) is also a typical Bohemian middle Cambrian species. The large, parabolic cephalon accounts for one third of the total body length, the thorax has 16 segments and the micropygeous pygidium has only one segment. The faintly indicated lateral ridges and the size and shape of the cephalon show that *Agraulos ceticephalus* was probably a species that burrowed in the ooze. The other species of this genus, known from middle Cambrian strata from the whole of Europe, presumably led a similar existence.

Solenopleura canaliculata (3), from Swedish middle Cambrian formations, is the typical species of the superfamily. It has a semicircular cephalon with a small, trapezoid glabella and — as distinct from the previous species — very pronounced palpebral lobes. It presumably also had good vision and lived on the firmer bed of shallow parts of the sea. Related species are common throughout the whole Anglo-Scandinavian part of Europe.

Bohemopyge discreta

Arthropoda
Trilobita

The majority of trilobites included in the superfamily Asaphacea are characterized by a large, wide and slightly convex carapace. The cephalon and the pygidium, which are both large, semiellipsoid or bluntly triangular in form, bear a close morphological resemblance to each other. The usually short thorax is composed of only 7—8 segments. Some species of asaphid trilobites were exceptionally large and, with their 50 cm length and roughly 25 cm width, were among the biggest trilobites in the Earth's geological history. The smooth carapace with only faint, shallow grooves separating the glabella from the cheeks, for example, the pygidial segments, the large, but flat, palpebral lobes and the nearly discoid cross section of the body confirm the hypothesis that these trilobites could swim, or more likely, that they floated just above the clay or marl bed of quiet parts of the sea. Although their first representatives appeared in late Cambrian seas, their main period was the Ordovician, and when it ended they died out. They are known chiefly from Europe. The biggest asaphid was undoubtedly *Opsimasaphus nobilis,* described by Barrande in middle Ordovician marl slates and marls from central Bohemia. Occasional, complete elliptical carapaces up to 45 cm long used to be found, and they were one of the chief Barrandian attractions for collectors. Related species are also known from middle and upper Ordovician strata in other parts of Europe.

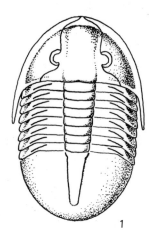

1

The geologically older species *Megalaspides dalecarlicus* (1) had a smaller, elongate carapace with a smooth, rounded or triangular cephalon and a flat pygidium with an only faintly defined, unsegmented central axis and pleurae. This species is characteristic of late Baltoscandian lower Ordovician positions; the genus as such is known throughout the whole of the early Ordovician.

Bohemopyge discreta (2) is related to *Megistaspis alienus* but has a morphologically more striking carapace, e.g. a well circumscribed glabella with shallow furrows and signs of segmentation of the pygidium. This species occur in Bohemia, but in the rest of Europe the genus *Bohemopyge* is known throughout all lower Ordovician strata.

Pricyclopyge binodosa

Arthropoda
Trilobita

The superfamily Cyclopygacea includes trilobites phylogenetically related to asaphid trilobites. Stratigraphically it also is restricted to the Ordovician, but the palaeogeographical distribution of its members is much greater, since they are known from North America and north Africa as well as from Europe. Cyclopygid trilobites are generally small, but have a characteristically constructed carapace. The greater part of the cephalon is accounted for by the glabella, which usually has one pair of oval glabellar lobes. Other striking features are the huge, faceted eyes, which are like an insect's eyes, i.e. compound, and are often joined together on the frontal and ventral aspect of the head. By contrast, the thorax is small and is composed of 5—6 segments. The triangular pygidium is also small. The mode of life of cyclopygid trilobites has long been a subject of erudite controversy. The huge, faceted eyes gave rise to the hypothesis that these trilobites lived at great depths and surfaced in the evening, the swollen glabella acting as a hydrostatic organ of balance during this migration.

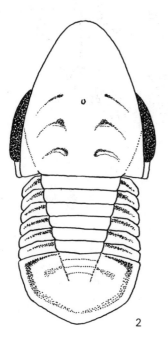

Pricyclopyge binodosa (1) is one of the largest and most typical species. It occurs in European lower Ordovician strata. Its carapace is up to 5 cm long, and on the median axis of the thorax it has two striking symmetrical grooves, which some researchers consider to be luminescent organs.

Novakella, formerly considered to be identical with the worldwide genus *Microparia,* is a stratigraphically important genus for dating Ordovician strata. It also comprises a relatively large species with an elongate, markedly curved glabella and a pygidium with only an indistinct or faint median axis, e.g. *Novakella bergeroni* (2).

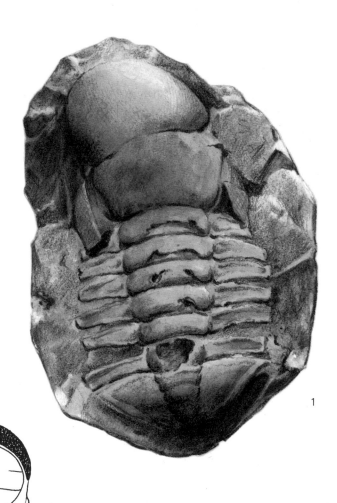

The rarer, but easily identifiable European Ordovician genus *Ellipsotaphrus* is notable chiefly for the structure of its eyes, which formed a continuous band across the whole frontal margin of the glabella, as in *Ellipsotaphrus monophthalmus* (3).

Spiniscutellum umbeliferum

Arthropoda
Trilobita

The family Thysanopeltidae is a characteristic group of moderately large trilobites whose innumerable species inhabited all the seas from the Ordovician to the Devonian period. The wide, semicircular cephalon has a distinctive glabella which widens anteriorly like a triangle and has three pairs of deep and intricate glabellar furrows. The large eyes are set in semicircular lobes. The flat, fan-shaped pygidium, with its extremely short median axis and long, fused ribs, is particularly striking. Thysanopeltids generally lived in shallow, well-oxygenated parts of the sea with gentle currents and a firm, calcareous bed, although they do not seem to have shunned the littoral zone and abundant remains are also found in coral reef sediments. They are known from Europe, North America, Africa, southeastern Asia and, more recently, South America.

Spiniscutellum umbeliferum (1) is a familiar thysanopeltid with an extremely rough-surfaced carapace. It is a typical fossil in fine dark, detrital, Bohemian limestones belonging to the very earliest part of the Devonian.

2

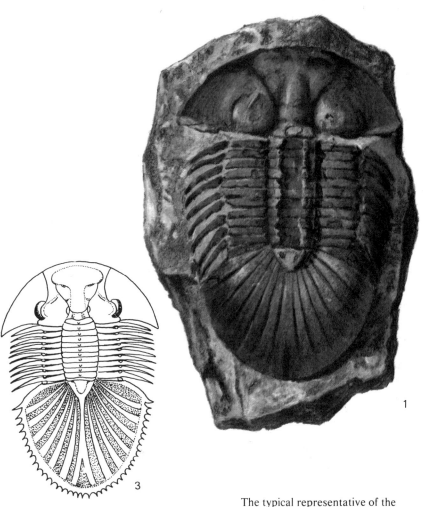

The similar, but stratigraphically somewhat younger, species *Bojoscutellum paliferum* (2) occurs in some parts of the Koněprusy, early Devonian, coral limestone massif in Bohemia. The remains of isolated pygidia sometimes form whole deposits in limestones. Some individuals — probably owing to the specific ecological conditions round the reef — attained a length of 30 cm and are thus ranked among the largest known trilobites.

The typical representative of the family, the genus *Thysanopeltis*, is geologically the youngest and is already fairly specialized. Unlike the other genera, whose members have a straight-edged pygidium, the pygidium of *Thysanopeltis* terminates in a spiny fan. A number of species are known from middle Devonian strata in Europe, North America and Asia. The species *Thysanopeltis speciosum* (3) is a rarer, but characteristic fossil of middle Devonian limestones in central and southern Europe.

Ectillaenus parabolinus

Arthropoda
Trilobita

The family Illaenidae is a worldwide group of trilobites with a moderately large, convex carapace. The cephalon and pygidium are almost the same size (isopygeous) and are morphologically so alike that in some species it was quite difficult to differentiate them. The wide, flat glabella is usually poorly defined, and the same applies to the median axis of the pygidium. Most species have small eyes, but pronounced palpebral lobes. The thorax, which accounts for one third of the length of the carapace, is composed of 8—10 segments. These trilobites were evidently not very good swimmers and spent most of their time crawling over the marl or sandy-argillaceous bed of the stiller, deeper parts of Ordovician and Silurian seas. Ordovician species are more abundant in Europe, Asia and north Africa; Silurian species are commoner in North America and southeastern Asia, while in European seas their number diminishes. At the end of the Silurian period they disappear altogether.

Ectillaenus parabolinus (1) is one of the best known members of the family. It has a relatively large, convex carapace up to 15 cm long, with a parabolically curved cephalon and an ellipsoidal pygidium. Similar species are known from Ordovician horizons all over Europe.

Bumastus, known from both Ordovician and Silurian formations, is a worldwide genus. *Bumastus hornyi* (2) frequented the firm tuffaceous sea bed round coral colonies in the shallow, Bohemian middle Silurian sea.

Another widely distributed genus is *Stenopareia*, which has a semicircular cephalon (3), a wide, relatively well distinguishable glabella and a smaller, but likewise semicircular pygidium (4). Various species of this genus lived in Europe, Asia and North America from the mid-Ordovician to the middle Silurian period.

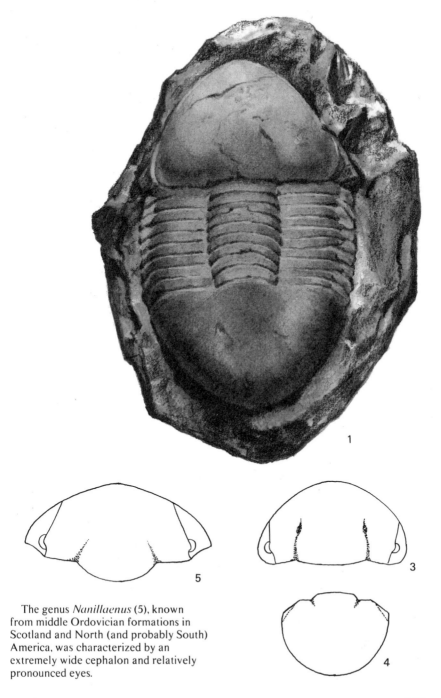

The genus *Nanillaenus* (5), known from middle Ordovician formations in Scotland and North (and probably South) America, was characterized by an extremely wide cephalon and relatively pronounced eyes.

Cornuproetus venustus

Arthropoda
Trilobita

The superfamily Proetacea is an unusually extensive group comprising several families of small, evolutionarily more modernly organized trilobites. Their carapace structure is extremely varied, as was their mode of life. They lived in the most diverse environments and formed both narrowly specialized local and endemic species and species with worldwide distribution. The oldest proetids appear in late Cambrian seas and the youngest disappear at the end of the Palaeozoic era, in the late Permian period; their extinction marks the end of the trilobites as a whole. Proetids usually have a semicircular or semi-ellipsoidal cephalon with a well-developed glabella and pronounced eyes of varying size set in the posterior half of the cephalon. Few blind species are known. The elongate thorax is composed of a large number of segments (up to 17) and the pygidium is also segmented. Geologically older proetids, especially Ordovician and Silurian, are commonest in Europe; from here they spread to the other continents during the Devonian period.

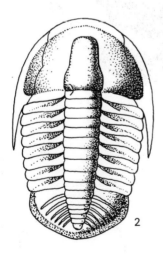
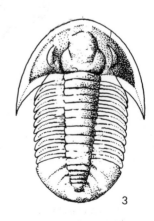

1

Cornuproetus, one of the commonest proetid genera, has a cephalon with a flat or concave rostral margin and long genal spines. A number of species are known first of all from European, Devonian strata (e.g. *Cornuproetus venustus* — 1).

Drevermania (2), an equally remarkable genus confined to Devonian and lower Carboniferous strata, is characterized by a wide cephalon, a small, bluntly triangular glabella and three pairs of glabellar furrows. Several European species are known; none have yet been found on other continents.

The members of the genus *Warburgella* (3) comprise some of the smallest, but stratigraphically extremely important trilobites. The remains of their carapaces measuring only a few millimetres, abound in lower Devonian limestones. Their occurrence is an excellent criterion for demarcation between Silurian and Devonian in the Hercynian palaeobioprovince, that is to say, in regions with unbroken sedimentation between the Silurian and the Devonian period (e.g. in central, southeastern and southern Europe, in north Africa and in central Asia).

Aulacopleura koninckii

Arthropoda
Trilobita

The large superfamily Proetacea includes a series of families and genera of small trilobites whose carapace is morphologically similar to that of actual proetids. The majority are known from middle Ordovician strata, but their peak periods were the Devonian and Carboniferous and the evolutionarily most modern types survived into the Permian period, i.e. to the end of the Palaeozoic era. Some of the most interesting of their innumerable known species are described below.

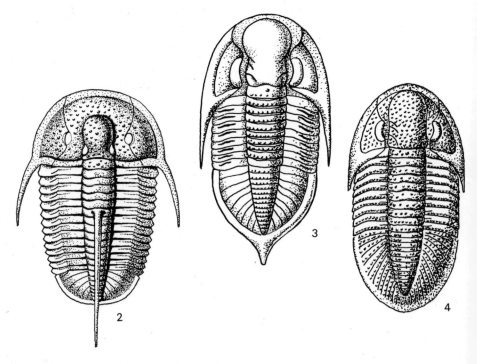

The minute species *Aulacopleura koninckii* (1) is typical of European, Silurian tuffaceous slates. Unlike other proetid trilobites of which only isolated intact specimens are known, it is not exceptional to find whole carapaces of this species in slates, and their remains often cover whole bedding planes.

Phylogenetically younger stages of these trilobites are also known. The genus itself is known from middle Ordovician strata in Europe, Greenland and Morocco and disappears in middle Devonian formations.

Otarion diffractum (2) is another interesting species. It is a member of

1

a cosmopolitan genus, which occurs in middle Ordovician to upper Devonian strata. The part found most frequently is the cephalon, with a half-oval, markedly convex glabella and small, highly protruding eyes. Whole specimens are very rare.

Remains of the carapaces of the characteristic genus *Weberides*, whose pygidium tapers to a long, sharp spike (terminal spine), are useful for the stratigraphy and correlation (i.e. reciprocal comparison) of European, upper Carboniferous limestones, especially in so-called paralic coalfields, where marine horizons alternate with the freshwater sediments of anthracite deposited in swamps. *Weberides mucronatus* (3) abounds in calcareous and argillaceous, upper Carboniferous sediments in northern Moravia and in Poland.

Phillipsia, whose rounded semi-ellipsoidal pygidium is larger that the cephalon (macropygeus), is a similar species. The axis of the pygidium is notable for its large number of segments (6—14). The typical species *Phillipsia gemmulifera* (4) is known in Europe, chiefly from English and Irish lower Carboniferous strata. Allied species occur in strata of the same age in North America and Asia.

119

Harpides grimmi

Arthropoda
Trilobita

The suborder Harpina contains interestingly constructed trilobites with a semicircular cephalon edged (especially in geologically younger species) with a flat pitted brim of variable width, shaped like a horseshoe. This curious arrangement of the head is associated with the animals' specialized mode of life, but the use of the cephalic fringe is uncertain. According to some researchers, such trilobites burrowed into the ooze on the sea bed and the border acted rather like a snow plough. Others are of the opinion that the fringe was a hydrostatic organ allowing the trilobites to swim or float. The latter hypothesis seems to be the more feasible, since the fringe was hollow (sometimes with a swollen anterior margin). Furthermore, Harpina also lived in shallow parts of the sea round coral reefs, where the bed was firm or rocky and the currents were strong; in consequence, the geographical distribution is often wide. On the other hand, there are local and sometimes blind species, which probably did actually burrow in the ooze. Further interesting features of Harpina include the simple eyes, the large number of narrow thoracic segments (up to 29) and the small pygidium. Representatives of the suborder are known from the late Cambrian; their large-scale development commenced in the Ordovician, attained its peak in the Silurian and Devonian, and the entire suborder died out at the end of the Devonian.

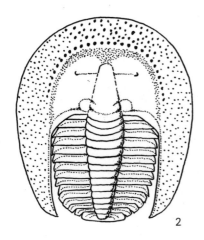

Members of the genus *Harpides*, known only from lower Ordovician strata, belong to the geologically oldest Harpina. Their carapace was up to 10 cm long and their cephalon had only flattened edges. As seen from *Harpides grimmi* (1), the typical border was not yet present.

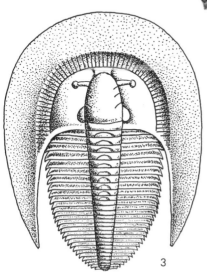

Eoharpes benignensis (2) already possessed a characteristically constructed carapace, however, and the cephalon had a relatively wide, regularly perforated fringe. This was a blind species and is found only in Ordovician slate layers. It seems to have been one of the Harpina which lived in the deeper, still parts of the Ordovician sea and burrowed in the ooze.

Bohemoharpes ungula (3) had a wide, hollow cephalic fringe tapering dorsally to long genal spines with a conical glabella and simple eyes on prominent lobes. This species, which is distributed over the whole of Europe, seems to have been able to swim.

Trinucleoides reussi

Arthropoda
Trilobita

The striking appearance of the specialized trilobites of the suborder Trinucleina and the ease with which they can be identified, even as fragments, make them popular both with collectors and with professional palaeontologists and stratigraphers. Their large, semicircular cephalon has a wide fringe riddled with small pits in a more or less regular pattern, which also goes round the cheeks on either side of the bulging glabella and terminates dorsally as sharp, narrow genal spines more than double the length of the body. The short, wide thorax is normally composed of 5—7 narrow segments and the pygidium is usually small and triangular. The eyes are generally small and simple; they are known only in young specimens, since they disappear later on, and adult individuals (apart from a few exceptions) were blind. Trinucleids are known from muddy and sandy sediments all over the world and so do not seem to have been restricted to a particular lithology. They appeared, flourished and declined during the Ordovician period, and only a minute proportion survived into the Silurian.

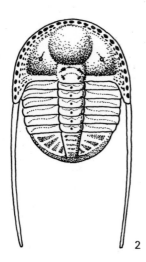

Trinucleoides reussi (1), from Bohemian middle Ordovician strata, was a member of a somewhat different, blind evolutionary branch of trinucleid trilobites. It differed from the rest by having an almost round body, a narrow, only finely perforated cephalic fringe and a strikingly prominent and almost conical glabella. It is an endemic species, i.e. it has been found nowhere outside Bohemia.

The numerous species of the genus *Tretaspis* (2), which was distributed over the whole of Europe and North America during the middle and late Ordovician and was recently found in north Africa, represent one of the main evolutionary lines of trinucleids. They are characterized by a high, anteriorly widened glabella and a narrow, but pronounced cephalic fringe.

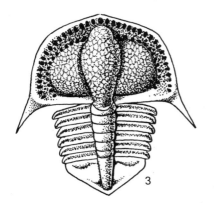

Trinucleid trilobites also occur in South America. For instance, in middle Ordovician strata in Argentina we find *Famatinolithus noticus* (3), which had an inflated glabella and cheeks, with a reticulate sculpture on their surface, a small thorax and a simple triangular pygidium. The massive distinctive cephalic fringe was only sparsely perforated.

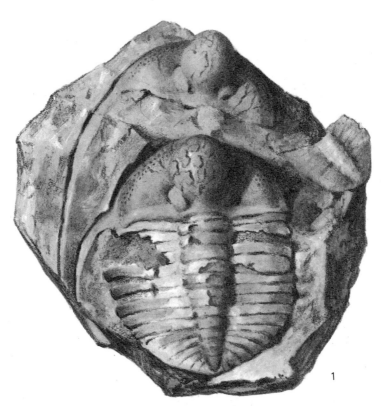

Flexicalymene incerta

Arthropoda
Trilobita

The order Phacopida is a very large group of exclusively post-Cambrian trilobites. Phacopids appeared during the early Ordovician and quickly achieved worldwide distribution. They are a pronounced component of biological communities in every type of Silurian and Devonian sea, but in the late Devonian the whole order degenerated, and by the end of the period it was extinct. In evolutionarily advanced types of phacopids the carapace is usually clearly divided into the basic parts — a cephalon, a segmented thorax and a relatively large pygidium. The glabella is generally marked with three pairs of furrows, and the striking, faceted eyes are seated on prominent semicircular (palpebral) lobes.

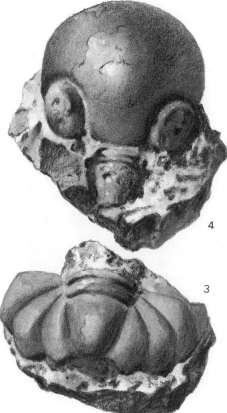

Flexicalymene incerta (1), from middle Ordovician strata in Bohemia, is a member of the phylogenetically old suborder Calymenina comprising specialized and morphologically very distinctive trilobites. It has a typical glabella, with three pairs of deep furrows, and a wide thoracic axis which forms an unbroken line with the wide pygidial axis. From the lower Ordovician to the Silurian period, *Flexicalymene* species were abundant all over the world.

The European Silurian species *Cheirurus insignis* (2) belongs to the typical suborder Cheirurina. Anteriorly, the characteristic, wide glabella overlaps the edge of the cephalon, the thoracic segments terminate in short, flat spines and the pygidium also tapers to more or less curved spines of varying lengths. *Cheirurus* and related genera are known in every part of the world.

The cheirurid trilobite *Sphaerexochus mirus*, which lived in shallow water round volcanic islands in the Bohemian Silurian seas, had a somewhat curiously constructed carapace. The parts found most frequently are sharp-contoured pygidia (3) and isolated, round, almost spherical glabellae (4), with a distinctive

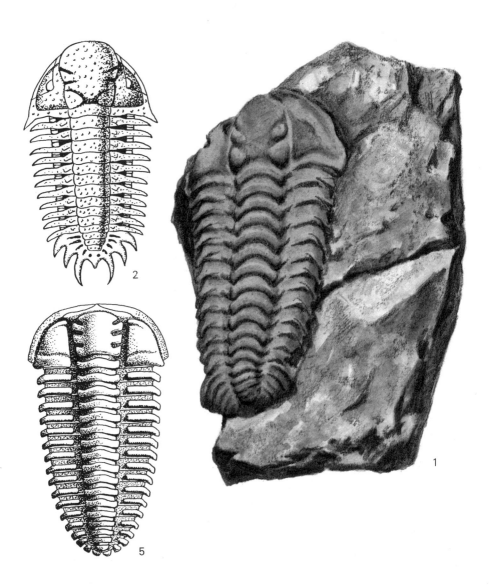

pair of posterior glabellar furrows. The furrows curve backwards to form round lobes and give the cephalon a vaguely skull-like appearance. A number of allied species are known from the rest of Europe and from Asia, North America and Australia.

The most striking features of the cheirurid trilobite *Placoparia barrandei* (5) are its well-defined glabellar furrows and its inflated thoracic segments. It is a representative of a genus whose species are known from middle Ordovician strata in Europe and Africa.

Phacops rana

Arthropoda
Trilobita

The family Phacopidae comprises evolutionally advanced trilobites whose innumerable species inhabited Silurian and Devonian seas all over the world. The majority have an elongate, oval carapace with a large cephalon; the pygidium also is relatively large, and the thorax is composed of 11 segments. The striking, bulging glabella widens frontally and usually stretches beyond the anterior margin of the cephalon. The compound eyes of phacopid trilobites have a distinctive structure characteristic of the entire family. In most trilobites the eyes are covered with a continuous transparent membrane, below which there are numerous small single elements known as ommatidia, which usually have lenses. There may be as many as 15,000 ommatidia in one compound eye. In phacopids the membrane is perforated, and in each small hole there is one, large ommatidium. The total size of the eyes and the number of ommatidia vary with the species. There are some lines in which the eyes grow steadily smaller, from species with a huge area of vision to completely blind types.

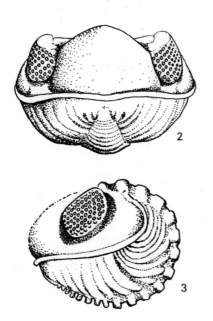

Phacops rana crassituberculata (1), from middle Devonian strata in the USA, is a typical phacopid. Whole specimens, with their carapace densely covered with large granules, are often found. This species, in common with many other phacopids, could roll up into a ball (sphaeroidal enrolment), the cephalon touching the pygidium (2 — anterior view; 3 — side view). This was chiefly a defence manoeuvre to protect the soft ventral part and to allow the animal to sink rapidly to the bottom of the sea if required.

The middle Devonian *Reedops cephalotes* (4) is a similar European species characterized by a wide, convex glabella extended forwards in such a way that it markedly modifies the contour of the cephalon. This and related North American species have strikingly large, bulging compound eyes (schizochroal).

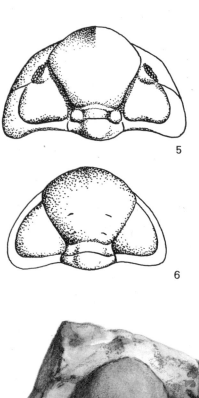
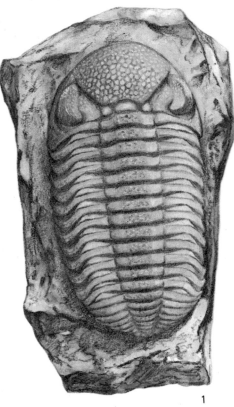
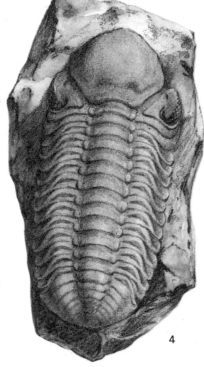

A noticeable reduction in the visual surface is exhibited in the later species *Cryphops cryptophthalmus* (5) from German upper Devonian formations, while the members of the specialized genus *Trimerocephalus* (6), belonging to the same evolutionary line, were completely blind. Reduction of the eyes was evidently related to a change in the phacopids' mode of life and possibly to their migration from shallow parts of the sea with flowing water to quiet creeks with a muddy bottom.

Odontochile hausmanni

Arthropoda
Trilobita

Members of the family Dalmanitidae likewise have a typical structure, are easily identified and are plentiful all over the world. As they successively invaded most marine environments, they occur in the most diverse sediment types, from the Ordovician to the Devonian. They all have an oval, well-proportioned body, a semicircular cephalon and a large pygidium with a long median axis which often terminates in a short, up-tilted (terminal axial) spine. The cephalon always has three striking pairs of glabellar furrows dividing the glabella into separate lobes, and usually very large, ellipsoid or hemispherical, prominent eyes. Because of their characteristic appearance and wide geographical distribution, most dalmanitid trilobites are excellent index fossils, especially for the Ordovician and Devonian periods, when the family as a whole had two peak development periods. In the Silurian period they tend to be less important.

The members of the genus *Odontochile* were important dalmanitid trilobites in the early and middle Devonian. They occur in Europe, North and South America, north Africa and Australia. It is interesting that plenty of isolated cephala and pygidia are found, but that whole

2

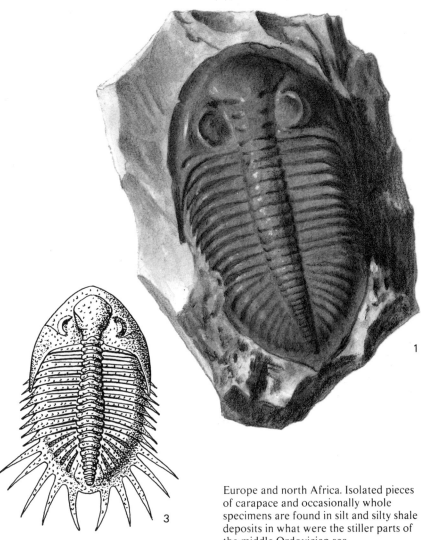

individuals are extremely rare. This complete specimen of *Odontochile hausmanni* (1) found in Bohemian middle Devonian limestones is thus an important find.

Dalmanitina proeva (2) is a stratigraphically very valuable dalmanitid species for the middle Ordovician cold water palaeoprovince of Europe and north Africa. Isolated pieces of carapace and occasionally whole specimens are found in silt and silty shale deposits in what were the stiller parts of the middle Ordovician sea.

Typical dalmanitid trilobites had a straight-edged pygidium with a narrow border, but in the members of the related subfamily Asteropyginae the posterior margin of the pygidium terminated in numerous spines. In the German middle Devonian species *Asteropyge punctata* (3) the spines were very long and radiated outwards from the posterior and postero-lateral margins of the pygidium.

Dicranopeltis scabra

Arthropoda
Trilobita

Early Palaeozoic seas were inhabited by large numbers of the members of the trilobite order Lichida, whose evolution produced huge forms which can only be described as bizarre. The order includes the largest known trilobites of all, whose carapace was over three quarters of a metre long. Lichids have a striking, wide glabella, divided by furrows into a number of separate lobes. The small eyes are crescent-shaped and are hard to distinguish. The free cheeks of the cephalon are narrow, but their margins are always extended to a long, curved spine. The individual segments of the thorax, also, terminate in rather long spines. The pygidium is usually flat, with a wide, but short median axis and three paired, foliate ribs, which again terminate in spines. The entire surface of the carapace is finely granular. Lichid trilobites appeared for the first time in the Ordovician period and spread rapidly over the whole of the world; they died out at the end of the Devonian period. The remains of their carapaces are found chiefly in limestones, so they probably frequented the hard beds of shallow, clear seas. Their wide, convex body shows that they had well-developed muscles and were able to crawl and swim. A problem still remains, however, regarding the purpose of their spiny exoskeleton, i.e. did the spines act as a hydrostatic organ of balance, or, more probably, were they only a support for crawling over the sea bed?

Dicranopeltis scabra (1) is a typical lichid trilobite; abundant isolated parts and occasionally whole carapaces are found in upper Ordovician and Silurian formations throughout Europe and in North America. The reconstruction of a whole exoskeleton (2), first undertaken by Barrande in 1852, shows the basic morphology of the carapace of this species in particular and of lichid trilobites in general.

One of the biggest trilobites ever known is *Terataspis grandis* (3), from middle Devonian strata in North America. The structure of the large pygidium, whose individual ribs are produced to extremely long, curved barbed spines, is especially noteworthy.

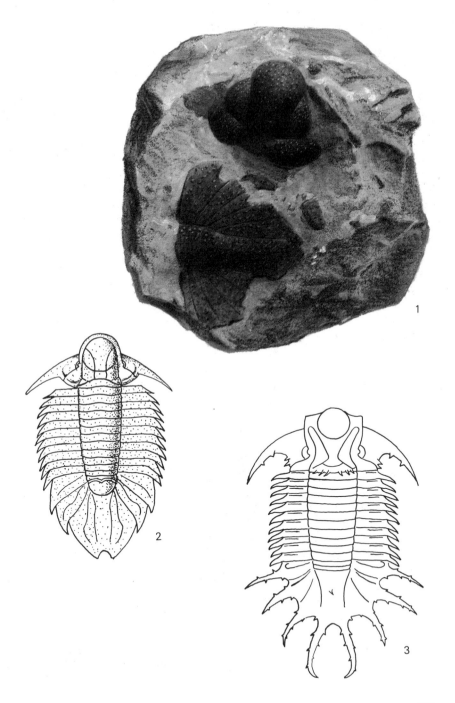

Selenopeltis buchi

Arthropoda
Trilobita

Perhaps the strangest forms among the trilobites are to be found in the order Odontopleurida, whose many species and individuals inhabited Cambrian, Ordovician, Silurian and Devonian seas. Their carapaces positively bristled with spines of every shape and size. There were spines not only on the margins of their carapaces, but also on the glabella, the cheeks, the palpebral lobes and the whole surface of the body. In some species the spines were straight, in others they were curved or branched, and sometimes the spines themselves bore spines. This curious organization of the carapace does not seem to have been altogether pointless. As in the case of the hairy larvae of some modern crustaceans, all this spiny beauty was presumably an organ of balance, enabling the trilobites to swim, or more likely for support on a soft muddy seafloor. The spines may also have been a passive defence against enemies — of which trilobites had more than enough, especially in Silurian-Devonian seas. The remains of odontopleurid trilobites occur mainly in fine sediments. Whole specimens are rare.

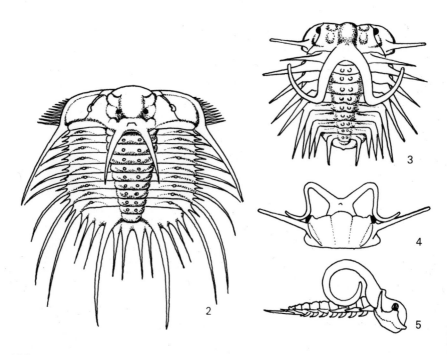

The geologically oldest odontopleurids include the large (up to 15 cm long) members of the genus *Selenopeltis* from Ordovician strata in Europe and north Africa, such as *Selenopeltis buchi* (1), which are characterized by a rounded trapezoid carapace with a large median axis. The cephalon, the body segments and the segments of the pygidium were all produced to flat and very long, recurved spines.

The typical representative of the whole order is *Odontopleura ovata* (2) from Bohemian middle Silurian strata, one of the not so large species which abounded throughout the European Silurian. Its wide, flat carapace had a border of fine spines of varying lengths and the median part of the occipital ring was produced into a pair of long occipital spines.

The Bohemian odontopleurid *Dicranurus monstrosus* (3), which was only about 5 cm long, was one of the geologically youngest and also, perhaps, the most remarkable. The marginal spines of the carapace pointed obliquely upwards and the occipital ring of the glabella was produced to two thick, bent and upcurved spines, rather like a ram's horns (4 — anterior view; 5 — side view). Related species are known from lower to middle Devonian formations in many other parts of Europe and in North America and Australia.

Meganeura sp.

Arthropoda
Insecta

Insects are ancient inhabitants of our planet. The oldest forms — small, wingless types, possibly related to the present-day springtails (Collembola) — appeared in Scottish Devonian formations. Winged insects first appeared in the Carboniferous period, and immediately attained large numbers. In addition to now extinct groups, they included members of orders which still exist today, such as mayflies (Ephemeroptera), dragonflies (Odonata), cockroaches (Blattaria) and others. During the Carboniferous period, insects developed rapidly and colonized the continents. At the end of the Carboniferous the character of their associations suddenly changed, and the majority of Carboniferous orders died out or changed significantly, to such an extent that the difference between the Carboniferous and Permian faunas is actually greater than the difference between the Permian and present-day insect faunas. Although the remains of Palaeozoic insects are known from all over the world, the palaeontological picture of their ancient associations is far from complete. Species living in tropical and subtropical lowlands and swamps overgrown with 'anthracite' forests had far better chances of preservation than species from dry foothills and mountains. A large number of fossil insects are already known, and their systematics are very complicated. We will therefore keep to the basic division of winged insects (Pterygota) into Palaeoptera and Neoptera. The former comprise insects with membranous wings not usually folded on their back. Palaeoptera abounded during the Carboniferous and Permian periods; the majority died out at the end of the Permian, and the only ones surviving today are Odonata and Ephemerida.

The dragonflies belonging to the genus *Meganeura,* which inhabited European Carboniferous 'anthracite' forests, include some of the biggest insects of all time, as they had a wing span of up to 70 cm. *Meganeura* remains (in particular the wings) are relatively rare, but occasionally a whole individual may be found (1).

Palaeodictyoptera was a purely Carboniferous-Permian and evolutionally primitive order. The genus *Ostrava,* from European early Carboniferous strata, is one of their oldest representatives. The wing of *Ostrava nigra* (2) shows both the main wing veins and remains of the dense network of small veins, the archaeodictyon. The remains of

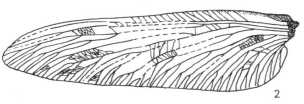

Carboniferous-Permian insects have sometimes been preserved in a remarkably perfect state. Finds in Bohemia and the USA are famous in this respect. The wings of *Bojoptera colorata* (3), from upper Carboniferous strata in central Bohemia, still show traces of their original colouring.

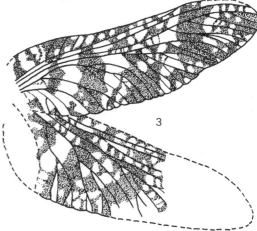

Phyloblatta sp.

Arthropoda
Insecta

The superclass Neoptera comprises more modernly organized insects which fold their wings over their abdomen when at rest. In some types the first pair of wings has hardened and has been converted to protective sheaths (elytra). As distinct from Palaeoptera larvae, which lived in water, the young larval stages of Neoptera lived on dry land. Like Palaeoptera, Neoptera appeared during the late Carboniferous period, but, except for cockroaches (Blattaria), they were still not very numerous. At the beginning of the Permian period, however, when the character of the insect faunas suddenly changed, they strongly predominated over palaeopterid forms, and, of the many Neoptera orders which evolved, most are still extant today. The majority of Palaeoptera lived in open spaces, because their inability to fold their wings onto their backs prevented them from taking shelter in confined spaces. This, together with the fact that they had aquatic larvae, was evidently the reason why this group declined so rapidly in the Permian period, when the damp climate quickly changed to a dry and arid type. The better adapted neopterid insects rapidly gained the upper hand and soon colonized the empty niches.

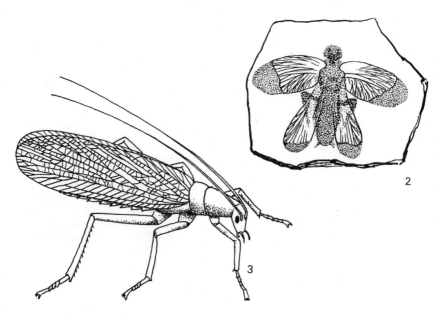

One of the most prolific representatives of Permian-Carboniferous neopterid insects were the cockroaches, whose innumerable species and individuals inhabited the undergrowth of the swamps and forests which ultimately gave rise to coal deposits. Most frequently we find their hard wing sheaths (= elytra) (the first pair of wings), which were washed into freshwater lakes and ponds. The wing shown here (1) belongs to the widespread genus *Phyloblatta*, which is known in Asia and North America as well as in Europe.

The European Carboniferous genus *Ettoblatina* had very wide wings with marked veining and a dense archaeodictyon. In rare cases almost complete specimens have been found, e.g. the species *Ettoblatina bohemica* (2).

The extensive, extinct order Protorthoptera, known from the late Carboniferous to the Jurassic, comprised insects in which cockroach (Blattaria), stone-fly (Plecoptera), grasshopper (Ensifera) and locust (Caelifera) characters were combined. *Sthenaropoda fischeri* (3) was one of its earliest representatives. Abundant finds in upper Carboniferous strata in France allowed exact reconstruction of these insects.

The wings of fossilized, late Permian scorpion flies of the species *Agetochorista tillyardi* (4) (order Mecoptera) still show traces of their original coloured spots. This species comes from the USSR, but other related Permian species are known from practically the whole of Europe, from inner Asia and from Australia.

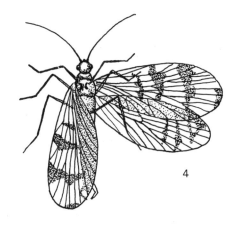

137

Acantherpestes vicinus

Arthropoda
Diplopoda

Among the interesting inhabitants of the damp Carboniferous-Permian forests were the millipedes (class Diploda). Although their segmented body was covered with chitinous armour, sometimes reinforced with calcium carbonate, so that it was capable of fossilization, there are far fewer fossil records of these creatures than of insects. The reason for this is their mode of life, e.g. under stones, in soil and humus and in layers of rotting vegetation. Problematical finds are already described from Silurian strata, but the first definite evidence of millipedes comes from the Carboniferous period, when several species were discovered. The late Carboniferous coal strata in central Bohemia and Carboniferous horizons in the state of Illinois in the USA are especially famous in this respect. Although some groups of millipedes have survived, practically unchanged, from the Carboniferous period to the present day, the majority of Palaeozoic Carboniferous-Permian forms were very different. Many of them were predacious and some were giants compared with recent types, as they were over 50 cm long.

One group of large millipedes, which lived only in upper Carboniferous forests, were the numerous members of the genus *Acantherpestes*, whose body segments were surmounted by long, branched spines. They are known from coals and from coal shales in Europe and north Africa. The typical species *Acantherpestes vicinus* (1), together with four other species, is fairly common in the Carboniferous strata of central and western European coalfields.

The members of the genus *Pleurojulus*, e.g. *Pleurojulus levis* (2), are another characteristic group of European upper Carboniferous millipedes. They had

1

a cylindrical body composed of over 60 smooth segments, a semicircular head and large faceted eyes.

One of the most widespread genera was *Archiscudderia*, comprising millipedes of moderate size with a small head, simple eyes and a smooth body composed of 15 or 16 segments. Their incidence is confined to the late Carboniferous. *Archiscudderia tapeta* (3) is the most abundant of the five species known and described in central Bohemia; similar species occur in the USA, e.g. in the famous Mazon Creek locality in Illinois.

3

Arachnocystites infaustus

Echinodermata
Cystoidea

Echinoderms (Echinodermata) are a major component of fossil associations. The majority have a strong calcareous skeleton (test) composed of regularly or (in primitive types) irregularly arranged plates, so they preserve well as fossils. Echinoderms occur only in marine deposits, and they are a very old group whose beginnings possibly date back to the Proterozoic era. By the beginning of the Palaeozoic era they were already differentiated into a series of morphologically and biologically independent groups. During the Palaeozoic the various groups produced a tremendous quantity of different forms, many of which, however, died out by the end of the era. Members of the primitive class Cystoidea are known virtually only from the Ordovician to the Devonian in marine sediments all over the world. They had a pouch-like or spherical case (theca) composed of porous polygonal plates. They were sessile and lived attached to the sea floor by a segmented stalk or by the base of the theca. The mouth was in the centre of the upper, ventral wall and from it led five radial grooves, often branched, which carried water containing microscopic food to the mouth; this is known as the ambulacral system. The ambulacral grooves sometimes continued from the theca to the free, segmented arms (brachioles). The excretory orifice and the genital orifice were localized eccentrically.

In Cystoidea of the order Rhombifera, represented by *Arachnocystites infaustus* (1), the pores on the thecal plates were arranged in a rhomboid pattern in which the rhomboids were connected by canals joining the corresponding pores on two adjacent plates. This species had a pouch-like theca, a short stalk and three long arms. It is an index fossil for sandy sediments of the middle Ordovician, cold water Mediterranean palaeoprovince of Europe, north Africa and southeastern Asia.

The theca of *Homocystites alter* (2) was composed of more or less regularly arranged, large, radially ribbed plates; it had a long stalk and numerous fine arms. Large, sparsely pored rhombs were distributed irregularly over the theca.

The similar genus *Cheirocrinus*, which had a regular goblet-shaped theca, a strikingly large excretory orifice on the side of the theca, numerous, thick arms and a long, massive stalk composed of circular segments, had a wide geographical distribution. The typical species *Cheirocrinus insignis* (3) is characteristic of English Ordovician strata.

Aristocystites bohemicus

Echinodermata
Cystoidea

As distinct from rhombiferous cystoids, which usually had a regularly constructed theca, members of the class Diploporita are noted more for their irregularity. Their theca is usually pouch-like, pear-shaped or spherical and is composed of large numbers of irregular, polygonal, massive calcareous plates, on which the pores are generally arranged in pairs (diplopores). Diploporite cystoids lived in the shallow parts of the seas, where there was a firm bed and well-oxygenated flowing water; they are often to be found near coral reefs. They were attached to the sea floor or to the body cases of other dead animals or to seaweed, etc. by the base of the theca, which was sometimes produced to a short stalk. Remains of their thecae and isolated plates, with the typical diplopores, are to be found in lower Ordovician to lower Devonian sand or chalk sediments in every part of the world. Since they are not very striking or beautiful, collectors are not particularly interested in them, although their scientific significance ought to earn them greater attention.

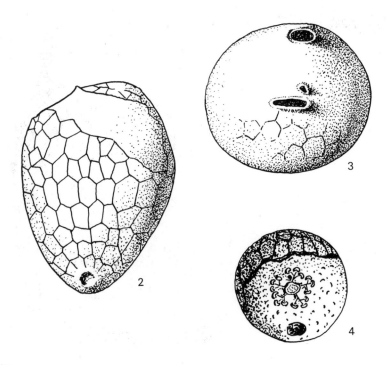

Aristocystites bohemicus (1), which has a large, pouch-like theca, is one of the best known species. The thecal plates are up to 1 cm thick and are easily distinguished in the rocks. Well preserved thecae and inner moulds (steinkerns) with plate impressions abound in middle Ordovician argillaceous-sandy and silt sediments in central and southern Europe, north Africa, inner and southeastern Asia and the eastern part of Australia.

The view of the under side of the theca of *Aristocystites bohemicus* shows a basal depression (2) at the spot where the animal was basally attached. The front view (3) shows the clearly discernible, fusiform central mouth at the apex of the theca, the small opening of the ambulacral system above it and the large excretory orifice on the upper margin.

The lower Devonian species *Proteocystites flavus,* which had a small, pear-shaped theca, is one of the geologically youngest diploporite cystoids. When viewed from above (4), the oral orifice can be seen at the apex of the theca, together with the short, radial ambulacral (food) grooves, which terminate in kidney-shaped facets at the places where the arms (brachioles) were attached. This species inhabited shallow, well-oxygenated parts of the warm lower Devonian sea.

Polydeltoideus plasovae

Echinodermata
Blastoidea

The class Blastoidea consists of specialized, extinct marine echinoderms with a bud-like, cup-shaped or spherical, regular calyx (theca) composed of 13 plates arranged in three rings: three basal plates, five radial plates and five deltoid upper plates. The radial plates are hollowed out to make room for the ambulacra, which are composed of small paired platelets and cover the ambulacral grooves. Deep below the ambulacral platelets there existed a complicated system of parallel canals known as hydrospires. The hydrospires were either open, as a row of parallel grooves running along the ambulacra (as in the order Fissiculata), or closed and opened on the apex of the theca through five pronounced openings known as spiracles (as in the order Spiraculata). This system had a respiratory function, transporting oxygenated water to all parts of the body. The excretory orifice opened on the apex of one of the deltoid plates.

Blastoidea were sessile and were attached by a stalk of varying lengths. They preferred shallow, well-lit and well-oxygenated, flowing water with a firm bed. The oldest rare, known species are from the Silurian period; in the Devonian their numbers increased, but the peak periods of the group were the Carboniferous and the Permian. At the end of the Permian period blastoids died out without leaving any descendants.

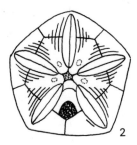

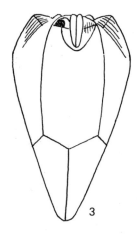

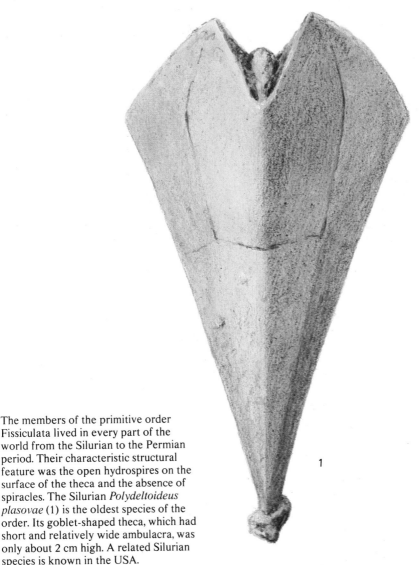

The members of the primitive order Fissiculata lived in every part of the world from the Silurian to the Permian period. Their characteristic structural feature was the open hydrospires on the surface of the theca and the absence of spiracles. The Silurian *Polydeltoideus plasovae* (1) is the oldest species of the order. Its goblet-shaped theca, which had short and relatively wide ambulacra, was only about 2 cm high. A related Silurian species is known in the USA.

Heteroschisma gracile (2 — summit view; 3 — side view) is a very interesting fissiculate blastoid with a tall, conical theca and pronounced, parallel hydrospire grooves running along the ambulacra. It is a member of a widespread and abundant genus in Devonian formations in the USA and Canada.

Pentremites godoni

Echinodermata
Blastoidea

The blastoids included in the order Spiraculata are organized on more complex and, it would seem, more modern lines than fissiculate blastoids. They are characterized by a specialized and highly efficient respiratory system. The hydrospire canals are sunk deep in the calyx and are covered by the side plates of the ambulacra; they open at the apex of the calyx as five simple or paired spiracles regularly distributed round the centrally situated mouth. Spiraculate blastoids are likewise known from the Silurian period, but they flourished chiefly in later Palaeozoic, i.e. Carboniferous and Permian, seas, where they produced a series of specialized and morphologically varied forms. Their remains are particularly abundant in calcareous rocks in every part of the world. Complete specimens with a stalk, and with rows of fine, unbranched arms (brachioles) bordering the individual ambulacra have been found.

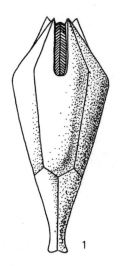

The oldest known representative of the spiraculate blastoids is the Silurian species *Troostocrinus reinwardti* (1) from Tennessee in the USA, which has a slender fusiform calyx about 3 cm high and short, narrow ambulacra.

Pentremites, whose many species occur in Carboniferous strata in North America, is the best known and most typical spiraculate genus. It has a typically built, squat, wide calyx shaped like a bud or a pear, wide, curved ambulacra and five, distinct, oval spiracles around the mouth. One of the most abundant and best preserved species is *Pentremites godoni* (2), whose individual growth stages are known from young specimens only a few millimetres high to adult animals measuring about 2 cm.

The morphology of the calyx of the Permian species *Deltoblastus timorensis* (3), which is known from the island of Timor and possibly from Sicily as well, and is one of the geologically youngest blastoids, is different. The calyx is ovoid and is truncated at the site of insertion of the stalk. The ambulacra run the whole length of the calyx right to its base. The spiracles are small and paired and open at the five apices of the ambulacra.

3

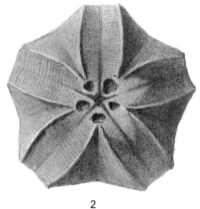

2

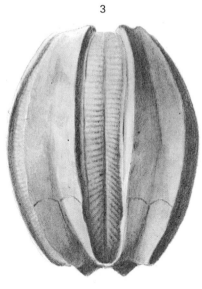

Lichenoides priscus

Echinodermata
Eocrinoidea

The class Eocrinoidea comprises completely extinct echinoderms whose incidence is confined to the Cambrian, Ordovician and Silurian periods. It is a largely artificial group and includes a variety of primitive types of echinoderms whose body is organized on similar lines to that of crinoids (sea lilies) and cystoids. Their skeleton was divided into a theca with arms (brachioles) and with a stalk or stem of varying lengths. In most known species the theca is pouch-like or goblet-shaped and consists of small, irregular plates. The brachioles are likewise primitive and are biserial (composed of two rows of platelets) and unbranched. The stalk, which is composed of narrow circular segments, is also more of the cystoid type. The pores in the sutures between the thecal plates is another cystoid feature. Eocrinoids all lived attached to the firm sandy bottom of shallow parts of the sea, although we also know species in which only the young animals had a stalk, and the adults swam about freely or simply rested on the thickened base of their theca. The most abundant finds of these echinoderms come from central and southwestern Europe, but they are also known in north Africa and the USA.

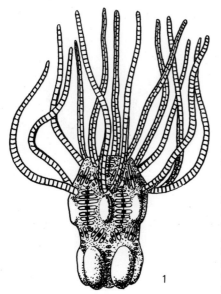

Lichenoides priscus (1, 2) is a distinctive and common Bohemian middle Cambrian species, with a comparatively regularly constructed theca composed of three rings of large plates (five in each ring) and possessing numerous long arms. The marked pores lie in regularly organized, deep grooves perpendicular to the sutures between adjacent thecal plates. Adult specimens were able to swim, as is evident from the large, oval, float-like cavities in the basal plates of the theca, as well as from the completely reduced stalk.

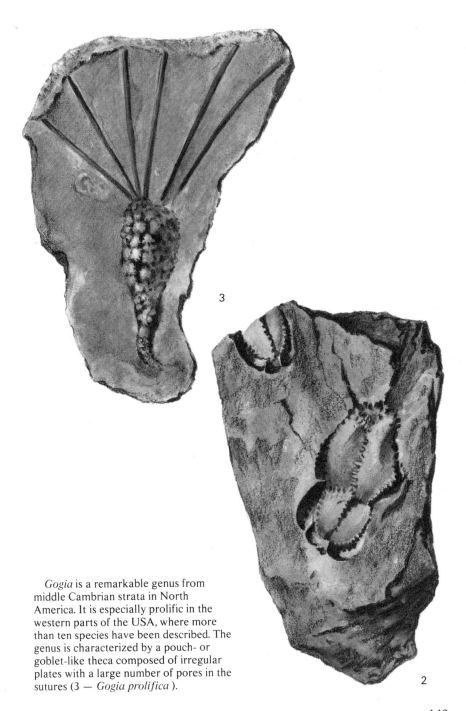

Gogia is a remarkable genus from middle Cambrian strata in North America. It is especially prolific in the western parts of the USA, where more than ten species have been described. The genus is characterized by a pouch- or goblet-like theca composed of irregular plates with a large number of pores in the sutures (3 — *Gogia prolifica*).

Scyphocrinites elegans

Echinodermata
Crinoidea

The ancient, palaeontologically important and morphologically rich class of the crinoids (Crinoidea), together with other groups of sessile echinoderms, has inhabited the seas since the beginning of the Palaeozoic era. Of all the echinoderms, crinoids are endowed with the greatest vitality and as 'living fossils' they are still encountered in the seas of today. Like all echinoderms, they are exclusively marine organisms. Their calcareous theca consists of a regularly constructed calyx with branched or simply segmented arms and usually has a stalk (column) — likewise segmented — of varying lengths, by which the animal is attached to the sea-floor. As a rule, the calyx and the arms form a radial pattern based on the number 5. The entire theca is extremely beautiful and is not unlike some exotic flower. Palaeozoic crinoids are not only excellent index fossils, but because they lived (and still live) in large associations, their remains became an important rock-forming component. Some coarse-grained Palaeozoic limestones consist of up to 95 % crinoidal detritus. The only drawback is that after they were dead, the bodies of crinoids broke up into single plates and stem segments (ossicles), making it difficult to identify them accurately.

Camerata are an entirely Palaeozoic order of crinoids, and one of the most important. They have a large cup formed by fusion of the basal parts of the arms as well as the actual calycal plates. Only the upper parts of the arms are free, but these are richly and minutely branched.

2

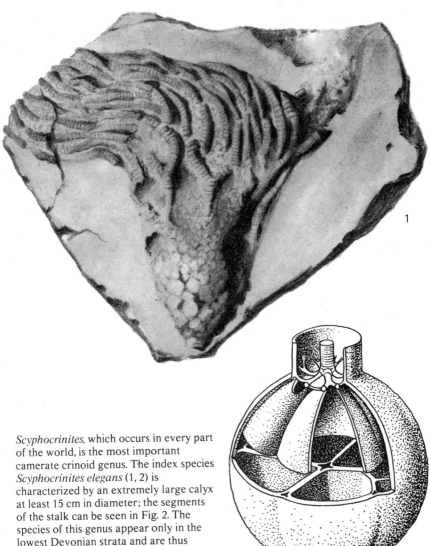

Scyphocrinites, which occurs in every part of the world, is the most important camerate crinoid genus. The index species *Scyphocrinites elegans* (1, 2) is characterized by an extremely large calyx at least 15 cm in diameter; the segments of the stalk can be seen in Fig. 2. The species of this genus appear only in the lowest Devonian strata and are thus excellent indexes of the division between the Silurian and Devonian periods. In some places the remains of *Scyphocrinites* are so abundant that they form horizons and lenticular masses several metres deep, known as scyphocrinite limestones.

The stalk of members of the genus *Scyphocrinites* terminated in a large, round float formed of small polygonal platelets and divided inside into several chambers (3). The float, called a lobolite, allowed these crinoids to drift with the ocean currents and explains how they encompassed the whole of the globe so quickly.

Caleidocrinus multiramus

Echinodermata
Crinoidea

The crinoids belonging to the order Inadunata are characterized by a regularly constructed calyx composed of firmly connected plates arranged in two or three rings, one above the other, in typical crinoid fashion. Inadunate crinoids have free, unfused arms composed of one or two rows of segments; they are usually long and fairly regularly branched. The stalk also is generally long, and its base is equipped with root-like processes which kept it firmly anchored in the sediment on the sea bed. Inadunata is another extinct group of crinoids which flourished in Palaeozoic seas, and their mode of life resembled that of most other crinoids of the same era. They tended to frequent the shallow parts of the sea where the water was well oxygenated, and specialized forms are even found in the breaker zone of coral reefs.

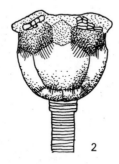

2

Caleidocrinus multiramus (1), from Ordovician sandstones and silt stones in central Bohemia, is one of the oldest typical inadunates. It has a tiny calyx, regularly branched, feathery arms and a long stalk with narrow circular segments. Allied species abound in Ordovician strata from the whole of Europe and in North America.

The worldwide Silurian genus *Crotalocrinites* was an interesting type. It had a spherical calyx the size of a fist and wide foliate arms formed by lateral fusion of the individual forked branches. The typical species *Crotalocrinites rugosus* (2) lived in shallow parts of the sea on the slopes of volcanic islands in Bohemia and England and on the island of Gotland in Scandinavia.

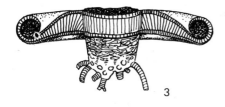

The Devonian species *Pernerocrinus paradoxus* (3) is an example of extreme adaptation to life in the breaker zone of coral reefs. It had a huge and unusually massive dish-like calyx up to 1.5 metres in diameter. The calycal plates are hard to distinguish; their number is generally secondarily augmented and they are irregular. All five arms are fused along their sides, giving rise to a flat or wide conical formation. These crinoids were most frequently anchored by root-like processes growing from the base of the cup, or by a massive stalk up to 20 cm wide and, in extreme cases, up to 5 m long. The geographical distribution of *Pernerocrinus* stretches from central Bohemia across southeastern Asia to Australia.

Taxocrinus coletti

Echinodermata
Crinoidea

Flexibilia is the last of the three crinoid orders; its origin, development and extinction took place during the Palaeozoic. Its members differ substantially from those of all the preceding groups in that their cup plates, and in some species the lower parts of the arms also, are loosely connected by flexible ligaments, so that they are capable of reciprocal movement. The lower parts of the arms are sometimes flexibly connected to the calyx, not immovably as in the case of camerate crinoids. The whole resembled a flexible bag bordered by the upper ends of the arms, and, as the living crinoid faced the direction of the current, it was well adapted to capture all the microscopic food borne towards it by the water (rheophilic). The calyx was attached by a long, flexible stalk composed of circular segments. Like other Palaeozoic crinoids, Flexibilia preferred shallow, well-lit parts of the sea with strong currents and a firm bed. It is interesting to note that, through the ages, crinoids migrated to the deeper parts of the sea, and the majority of present-day sessile species are found at depths of several thousand metres, while only modern species with stalkless, non-sessile adults live in the shallow and littoral zones of the sea.

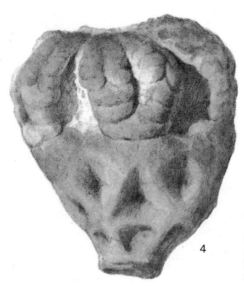

4

Taxocrinus, the characteristic genus of the flexible crinoids, has a small calyx and large, multiple-branched arms. Numerous species of this genus lived in the Devonian and lower Carboniferous seas of Europe and particularly North America. *Taxocrinus coletti* (1, 2) is an important species in lower Carboniferous marly limestones in the USA.

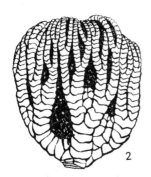

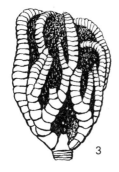

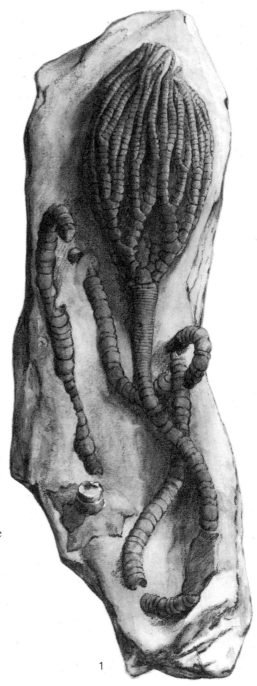

Protaxocrinus, with the typical species *Protaxocrinus svobodai* (3), is a related, but geologically older genus. The species in question was relatively abundant on the shallow limestone bed of the sea round Silurian volcanic islands in central Bohemia. The diagram of the crown shows a complete specimen seen from behind, with the typical oblong anal plate incorporated into the radial pattern.

The genus *Pycnosaccus,* whose many species are known from Silurian and Devonian strata in Europe and North America, was characterized by a spherical, swollen calyx, composed of large plates with radial ribs on their surface and of relatively short, forked arms. *Pycnosaccus bucephalus* (4) is typical of European Silurian tuffaceous limestones.

Stromatocystites pentangularis

Echinodermata
Edrioasteroidea

In some Palaeozoic sediments are found large numbers of small thecae composed of small, overlapping plates and attached to the substrate by the whole of their flat under (aboral) surface. The mouth is in the centre of the upper (adoral) surface and forms the centre of a star composed of five, straight or curved ambulacra covered with two rows of small interlocking platelets. The anal orifice (periproct) which is usually covered by a squat pyramid of triangular platelets, is localized eccentrically. These thecae belong to the remarkable echinoderms of the class Edrioasteroidea, which lived in shallow parts of the sea, attached to the firm bed, or, very often, to the exoskeletons of other animals. The spiny trilobites of the genus *Selenopeltis* were particularly frequent hosts of edrioasteroids, which carried them to fresh sources of food. Edrioasteroidea is an ancient class of echinoderms known all over the world from the early Cambrian to the Carboniferous, but similar forms (Tribrachidium) have actually been discovered in Precambrian sediments.

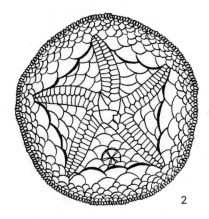

2

Stromatocystites is one of the oldest edrioasteroid genera. Most of its species have been found in Europe, but some have also been found in Newfoundland. *Stromatocystites pentangularis* (1), from middle Cambrian sandstones in Bohemia, is a typical example.

The cumulative genus 'Hemicystites' (2) is typical of Ordovician horizons. According to the latest studies, this name covers several, related, but separate genera. These types of edrioasteroids have a discoid theca measuring 1 — 2 cm, composed of overlapping plates and with short, thick and slightly curved

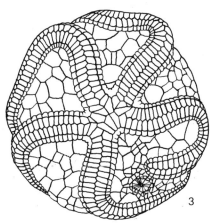

ambulacra. They have so far been found in Europe, North America and Australia.

Edrioaster bigsbyi (3) is a representative of more intricately organized species than the ones described above. Its rounded, flexible theca is composed of comparatively large plates, and it has strikingly curved, wide ambulacra stretching to the under (aboral) side of the theca. This and several related species are fairly abundant in North American Ordovician strata and somewhat doubtful finds have been made in Europe.

Trochocystites bohemicus

Echinodermata
Homalozoa

As a result of recent intensive study of echinoderms known as 'carpoids', views on the whole of this remarkable group have undergone a radical change which has led to their being divided into the subphylum Homalozoa with some being assigned to the subphylum Calcichordata. The latter are organisms much higher up the evolutionary ladder and closer to vertebrates. Homalozoans were echinoderms with a rounded or pouch-like theca composed of large numbers of calcareous plates and were capable of some free movement. On the superior face of the theca there was a mouth with a short ambulacrum which terminated in a segmented arm (brachiole) of varying length. At the end of the theca was a hollow, segmented stalk which tapered off to a sharp-pointed tip. In homalozoans the stalk was used for locomotion, however, and not for anchorage. Homalozoans mostly frequented the still, deep parts of the sea, where they burrowed into the ooze or crawled on the sea bed; some species probably only lay on the sea bed. The most frequent finds are made, therefore, in clay and marl sediments, in every part of the world; their ages range from Cambrian to middle Devonian, with the maximum incidence in the Ordovician period.

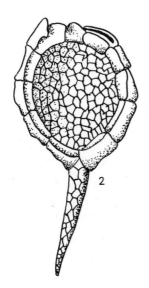

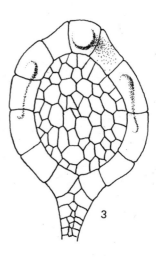

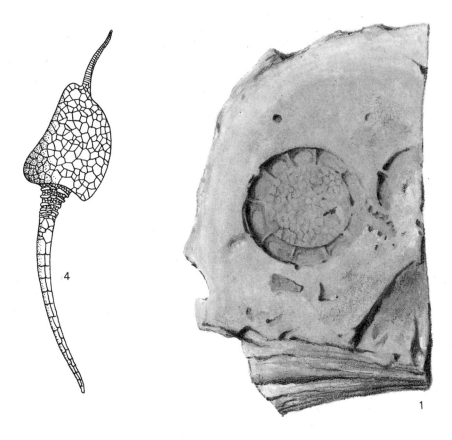

Trochocystites bohemicus (1, 2), found in middle Cambrian strata in Bohemia, Spain and France, had a round, discoid theca about 2 cm across and a short stalk with two or three rows of plates. The theca is morphologically differentiated, the margins being formed of large, rectangular plates and the centre of small polygonal plates. These homalozoans had no brachiole, but just two short ambulacral grooves leading from either side of the mouth in the superior face of the theca.

The similar French Cambrian species *Gyrocystis barrandei* differed from *Trochocystites* in respect of its ovoid theca, which had sharp margins and a slit-like mouth with no ambulacral grooves (3).

Dendrocystites barrandei (4), from middle Ordovician silt stones in Bohemia, is an example of homalozoans with a pouch-like theca. On the superior face there was a wide, curved brachiole formed of two rows of thin platelets. The conical upper part of the massive stalk was composed of transverse rows of small platelets and the lower part of larger semicylindrical plates. Possible related species are known from the Baltic-Scandinavian region and from Scotland.

Archegonaster pentagonus

Echinodermata
Somasteroidea

The subclass Somasteroidea comprises echinoderms capable of free movement, which some time ago supplied specialists with a minor sensation. They had always been regarded as an exclusively Palaeozoic group whose last members died out in the Devonian period. In 1952, however, a living representative of this subclass — which is over 250 million years old — was described and the 'animate fossil' was not so very different from its long extinct ancestors. Somasteroidea is a highly specialized group of echinoderms resembling starfish (Asteroidea), but whereas in true starfish and brittle stars (Ophiuroidea) the strongest body axes are the ambulacral axes down the middle of the arms, while the margins of the theca are flexible, in somasteroids the reverse applied, i.e. the ambulacra radiating from the central mouth were flexible and the margin of the theca was reinforced with massive rectangular plates known as marginals. In addition, parallel rows of rod-like platelets known as virgals, which provided further stiffening, led from the marginals to the ambulacra.

It was always supposed that somasteroid echinoderms preferred still water, where they burrowed into the soft ooze and lived on organic substances floating just above the sea bed. This hypothesis was confirmed by recent members of the subclass found in the Gulf of Mexico, where they lived in sandy-argillaceous beds at a depth of only 3—5 m. Palaeozoic somasteroids are known from Ordovician strata in Bohemia, France and Spain, from Silurian strata in Australia and from Devonian strata in the USA, so palaeogeographically they were widely distributed.

Archegonaster pentagonus (1), found in Bohemian lower Ordovician strata and one of the oldest known representatives of the subclass, had a flexible pentagonal theca with relatively small, convex marginal plates. Fragments of thecae and occasionally whole individuals are found chiefly in siliceous (originally calcareous) nodules from Llanvirn shales near Prague and Rokycany (Czechoslovakia).

The diagram (2) shows structural details of the arm of the theca of the genus *Archegonaster*, with the wide ambulacral field, the marginals and the virgal platelets.

Villebrunaster thorali, from French lower Ordovician strata, had a somewhat different structure. The theca was wide and star-shaped, the marginals were rudimentary and their function as stiffeners was almost completely taken over by the strongly developed virgals. Fig. 3 is a diagram of the structure of one of the arms.

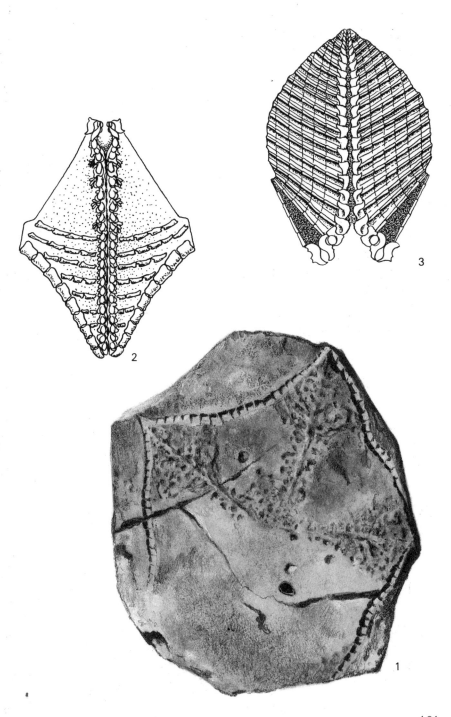

Encrinaster roemeri

Echinodermata
Ophiuroidea

The brittle stars (Ophiuroidea), an ancient subclass whose members are still an important component of marine associations, belong to the group of echinoderms capable of active movement. They are encountered in all the seas with the exception of inland seas and gulfs, where the salinity of the water is too low, e.g. the Baltic. They live at every possible depth — under stones in the littoral zone and at depths of thousands of metres. They are easily distinguished from true starfish by their flat (usually circular or pentagonal) body disc, which is sharply divided off from the long, thin arms; the latter are round in cross section and grow narrower until they terminate in a sharp-pointed tip. The arms are very flexible; they are composed of articulating segments and enable the animal to crawl or swim with considerable agility. Brittle stars usually live on microscopic food and on organic substances on the sea bed. They are divided into two orders, according to the structure of the body disc and the ambulacral fields. The members of the first, more primitive order Ophiurida appeared in Ordovician period while the more modern Euryalae lived in the Carboniferous period. Both groups of brittle stars have survived to the present day in all seas, the fossil species, however, being not particularly common.

1

Fine, early Devonian Hunsrück shales in Germany have furnished large numbers of beautifully preserved fossil remains, including *Encrinaster roemeri* (1), with a pentagonal body disc and long, fusiform, narrow-tipped arms. The genus *Encrinaster* is known from upper Ordovician and early Carboniferous strata in England and Scotland, as well as from central Europe.

The diagram of the lower part of the theca of the genus *Encrinaster* (2) shows a fusiform arm with the ambulacra, rows of openings for the pseudopodia and part of the central disc circumscribed by marginal plates bearing small spines.

The genus *Furcaster*, with its subtle body structure and long, slender arms, is widely distributed. The pronounced adoral plates forming a five-pointed star are typical. *Furcaster palaeozoicus* (3) is relatively common in European early Devonian strata; allied Ordovician to early Carboniferous species occur in North America and Australia.

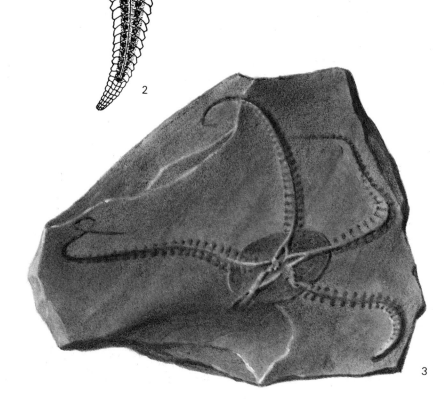

Siluraster perfectus

Echinodermata
Asteroidea

Starfish (subclass Asteroidea) are familiar echinoderms still encountered today, although many of their species date back to the early Palaeozoic. They are capable of free movement and have a flat or convex body disc, which is not sharply circumscribed, but merges without a break into the wide, flat, radial arms. There are usually five arms, but there may be as many as 45. From the mouth, which is on the under side of the disc, ambulacra radiate out to the individual arms; in these freely moving forms they have lost their nutritional function and have become locomotor organs. Along each ambulacral groove runs one branch of the animal's water vascular system and from it, through openings in the ambulacral plates, extend rows of pseudopodia (tube feet) fitted with suction discs, which enable the starfish to crawl. Unlike brittle stars, starfish are predacious. They can thrust out their muscular stomach through their mouth to engulf food and then digest it outside their body; they are even able to open mollusc shells. Starfish first appeared during the early Ordovician period and flourished during the Palaeozoic era. Towards the end of the Palaeozoic some lines died out, but others survived into the Mesozoic, when the whole class revived again. Starfish are still an important component of marine fauna and can be regarded as living fossils which have remained substantially the same over a period of 480 million years.

In middle Ordovician silty and sandy slates in Bohemia and England are found fragments and occasionally whole discs of the massively built, long- but slender-armed members of the genus *Siluraster* (1), whose body measures about 5 cm across.

The pretty starfish of the genus *Protopalaeaster* are typical of lower Ordovician formations in North America, especially the USA, and are also known from Ordovician strata in Europe and inner Asia. The diagrammatic reconstruction of the American species *Protopalaeaster narrawayi* shows the oral side of the disc, with the central mouth and the ambulacral system (2), and the aboral side, with a massive central area composed of regularly organized plates and with short, round-tipped arms, also of regular rows of plates (3).

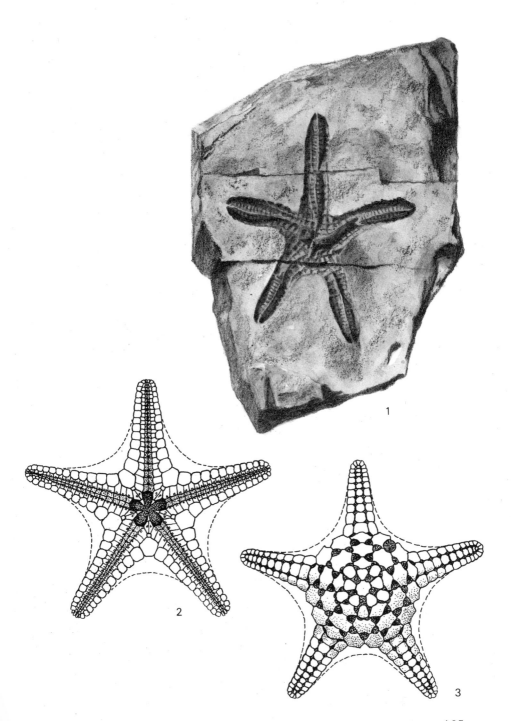

Rhenechinus hopstätteri

Echinodermata
Echinoidea

Sea urchins (Echinoidea) are probably the most popular class of echinoderms — not only among specialists, i.e. palaeontologists and zoologists, but also with holiday-makers, who on rocky shores with shallow water sometimes look in vain for a spot where their bare feet will be safe from the spiny bodies. Anyone who happens to sit down on a sea urchin will remember the experience for a long time. Sea urchins are highly organized echinoderms with a regularly constructed test, the surface of which is covered with articulating, movable spines of various shapes and sizes. The test is generally hemispherical or flat and discoid; five ambulacra composed of porous plates and five interambulacra formed of solid plates can be distinguished. The pseudopodia of the water vascular system — i.e. locomotor and respiratory — pass through the pores in the ambulacral plates. The mouth lies on the oral (under) side of the test and the anal orifice (periproct) is at the apex of the aboral (upper) side. Sea urchins live in all seas with a normal degree of salinity and at diverse depths. They generally crawl on the sea bed, but some species make quite deep burrows in the sand. Others attach themselves to rocks and scarcely move at all. They live on organic débris and algae and are sometimes predacious. Sea urchins did not play a very important role in Palaeozoic seas. Their period of prosperity began in the Mesozoic and Cainozoic eras, and they are still a flourishing class today.

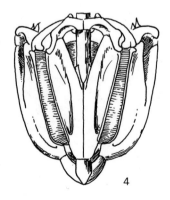
4

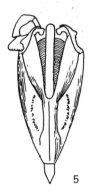
5

Hunsrück Devonian slates in Germany have a beautifully preserved and very rich fauna. Almost complete tests of the sea urchin *Rhenechinus hopstätteri*, composed of small polygonal plates with delicate spines and with finely constructed ambulacra have been found. The individual with intact calcareous jaws in life position (1) is a unique find. Fig. 2 shows a detail of the masticatory apparatus of the genus *Rhenechinus* and Fig. 3 a detail of an ambulacra complete with pores.

Isolated test plates and parts of the masticatory apparatus (4) — known as Aristotle's lanterns — of large sea urchins of the genus *Lepidocentrus* are sometimes washed and weathered out of European and North American Devonian limestones. The 'lantern' is a very complex, but practically constructed organ. It is shaped like a five-sided pyramid with the point downwards and is composed of large mandibular plates with teeth on their apex (5). The jaws could be opened and clamped together and probably functioned like a grab.

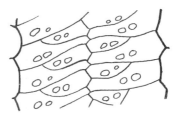

Koremagraptus spectabilis

Hemichordata
Dendroidea

The marine animals of the order Dendroidea stand between Achordata and Chordata — animals without and with a notochord, i.e. a skeletal rod. They formed dendritic colonies known as rhabdosomes, composed of many individuals interconnected by stalks. Each individual had its own well-developed nervous-stolon system and a solid body axis (pectocaulus). It was encased in a tubular theca composed, like the whole rhabdosome, of a hard substance related to chitin. These animals were consequently well equipped for fossilization. Dendroids were generally sessile and lived in shallow water near the shore, either directly on the rocky bed or, possibly, attached to large stalks of littoral algae. Occasionally, some species took to an epiplanktonic existence, i.e. they attached themselves to floating objects and let themselves drift. Dendroidea occur only in the Palaeozoic era and attain maximum incidence in the Silurian. In Devonian strata there is a marked drop in the number of dendroids, and as the Carboniferous ends their last representatives disappear altogether.

1

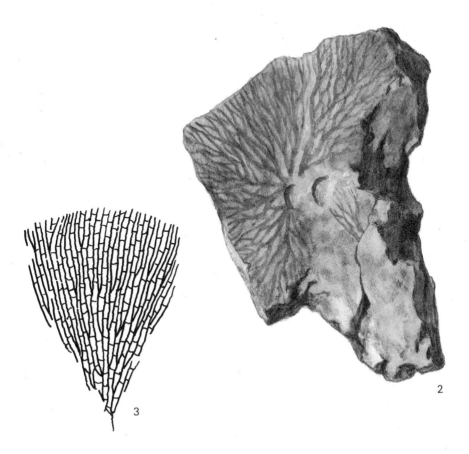

Koremagraptus spectabilis (1) is a striking dendroid with a large, tuft-like or funnel-shaped rhabdosome composed of irregularly interwoven and interconnected branches bearing large numbers of thorny processes formed from the tangled free ends of the long thecae. Different species of the genus *Koremagraptus* occur in Ordovician, Silurian and Devonian strata throughout Europe. *Thallograptus* was a widespread genus, both stratigraphically, from Ordovician to Devonian formations, and geographically. Its robust rhabdosomes, with numerous, short branches each carrying a quantity of long, narrow thecae, have been found in Europe, North America and, quite recently, in Asia. *Thallograptus muscosus* (2) is the commonest of the more than 15 species found in European Silurian strata.

Dictyonema flabelliforme (3) is a stratigraphically important species with a delicate rhabdosome composed of forking branches interconnected by cross-supports (dissepiments). Its incidence is confined to the Ordovician period. It belongs to those epiplanktonic species which could be carried long distances by the ocean currents. It has been found in Europe, north Africa and Asia, but allied species are known from upper Cambrian to early Carboniferous strata all over the world.

Monograptus priodon

Hemichordata
Graptoloidea

Over 150 years ago, strange remains reminiscent of gleaming threads or pencil lines were found in Ordovician and Silurian black slates. They were thought to be a freak of nature and were called graptolites, from the Greek *graptos*, meaning written on, and *lithos*, meaning stone. It was not until long afterwards that these 'written-on-stones' were identified as the remains of one of the most important fossils of all — evolutionarily highly organized and extremely valuable for biostratigraphy (strata dating). Graptolites were marine animals which, as distinct from dendroids, formed simple and usually long chitinous colonies (rhabdosomes) like wands. The animals were attached to one or to both sides of the axis of the stipe. The whole colonies generally lived freely like plankton, suspended from their own floats, or like epiplankton. This explains the worldwide distribution of many of their genera and the fact that they are found in large quantities in fine, black shales and slates rich in pyrite (graptolitic shales), which represent sediments deposited in low-energy, poorly aerated environments. In the geologically short time during which graptolites are known, i.e. from the Ordovician to the early Devonian, the group developed with explosive rapidity, which is another reason for its exceptional stratigraphic value. Practically every separate horizon contains a different association, and this allows not only the reciprocal comparison of layers, even on remote continents, but also the reconstruction of palaeogeographical relationships between the world graptolite provinces.

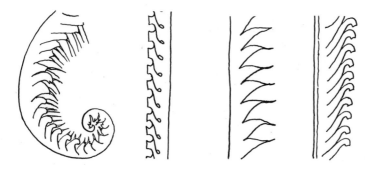

4

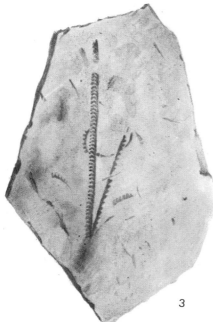

Monograptus priodon (1, 2) with a long, straight stipe (= rhabdosome) is one of the best known graptolites. The thecae are arranged one above the other on one side of the stipe and are characterized by a hook-like mouth. This species abounds in Silurian strata in every part of the world, and occasionally occurs in limestones.

The European and North American Silurian *Monograptus lobiferus* (3) is another distinctive species. It has slender, finely built rhabdosomes and the thecae have a rounded, jutting mouth.

The morphology of the rhabdosomes of the various graptolite species and genera is very diverse (4) and it is still the main criterion for their systematic classification.

Mitrocystites mitra

Calcichordata
Mitrata

There is scarcely a group of Palaeozoic animals which, in recent years, has attracted so much attention and been the source of so many controversies as the subphylum Calcichordata. Until recently, these animals were ranked among the carpoid echinoderms (see earlier), but in a detailed study of their anatomy the English researcher R.P.S. Jefferies found that their body organization corresponded to that of much higher animals related to the chordates. The body of calcichordates resembles the body of carpoids, but the likeness is merely a morphological similarity of two different groups of organisms leading a similar existence. Calcichordates had an intricately organized body, with a digestive system, a brain centre with optic nerves and a gill respiratory system, etc. Their organ of locomotion was the flexible segmented 'stalk' on the dorsal wall of the theca. Calcichordates preferred the still, deep parts of the sea with a muddy bed. Some species burrowed in the ooze or crawled or slithered on its surface, while others were able, perhaps, to swim a short distance above the ocean bottom. The first known calcichordates came from Cambrian strata; they attained their maximum incidence during the Ordovician period, and their last known representatives are of middle Devonian origin. Their numerous species are scattered all over the world, but the majority are local, only a few achieving wide geographical distribution.

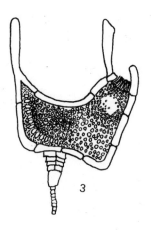

3

The best known mitrate species is *Mitrocystites mitra;* it has an oval theca, whose superior face is formed of small platelets and the inferior face of five large plates (1). The margins of the theca are reinforced with large, long marginal plates stretching from the upper to the under side of the theca. The stalk, composed of two rows of platelets, is only a little longer than the theca. This species is characteristic of middle Ordovician strata in Bohemia, France and Spain.

As its name implies, the Bohemian middle Ordovician species *Lagynocystis pyramidalis* (2) has a high, pyramidal theca mainly composed of large plates, with a long, widely conical stalk on its base. It is notable for the short process at the apex of the theca, which covers the mouth.

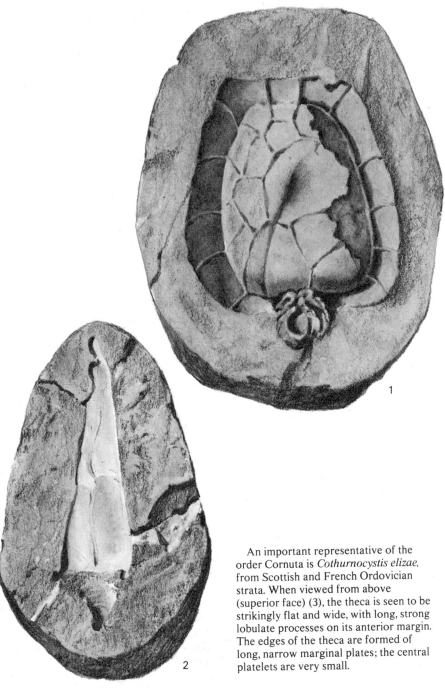

An important representative of the order Cornuta is *Cothurnocystis elizae*, from Scottish and French Ordovician strata. When viewed from above (superior face) (3), the theca is seen to be strikingly flat and wide, with long, strong lobulate processes on its anterior margin. The edges of the theca are formed of long, narrow marginal plates; the central platelets are very small.

Eucephalaspis lyelli

Chordata
Agnatha

The Palaeozoic seas did not give birth only to manifold classes of invertebrate animals. From this era we already know remains of chordate animals (Chordata), which had a skeletal rod, i.e. a notochord, and a more or less ossified internal skeleton. The oldest known true chordate remains are today described as late Cambrian, showing that they are just as old as the other animal phyla and that we must look for their origin in Proterozoic times. The Palaeozoic members of the class Agnatha, whose name means jawless, are a curiously formed group of 'fish-like vertebrates'. Their internal skeleton was evidently represented only by the notochord, but they had well-developed external armour made of solid bony plates. In addition to the two frontal eyes on their semicircular cephalon, they had a third, parietal eye, but only one nostril. They had no paired fins and no true jaws, and the mouth was a plain orifice. Rare fragments of the armour of these primitive vertebrates are already known from the late Cambrian and the Ordovician period, but their peak period was the Silurian, and they died out before the end of the Devonian. Judging by their massively built, usually flat-sided body and their comparatively small, irregular caudal fin, the majority do not seem to have been very good swimmers. They probably lived near the bottom in quiet and not very deep parts of the sea. Some may have lived in estuaries and rivers.

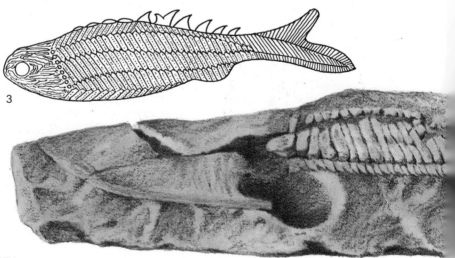

3

Eucephalaspis, whose many species are known from Silurian and Devonian strata in Europe, Spitzbergen, North America and eastern Asia, is one of the most prolific genera. As a rule, only the flat, semicircular head plates, whose posterior margins are extended to form sharply pointed spines, are found; the rectangular scales covering the body are found less frequently (1, 2 — *Eucephalaspis lyelli*).

The members of the European genus *Birkenia* (3) are relatively lightly built. They have an extremely concave caudal fin, showing that they were quite good swimmers. Whole specimens about 10 cm long have been found in Silurian and lower Devonian layers, especially in Scotland and Scandinavia.

Pteraspis cornutus (4), only 6 cm long, is another representative of the class Agnatha. Remains of species of the genus *Pteraspis* have been found in lower Devonian strata in Europe, Spitzbergen, North America and Siberia.

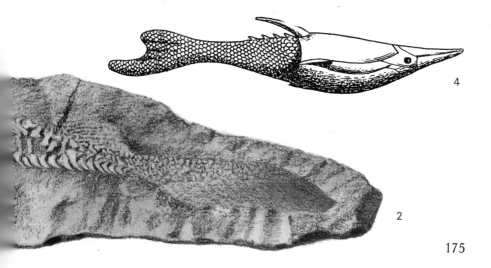

Bothryolepis canadensis

Chordata
Placodermi

While the class Agnatha developed mainly during the Silurian period, Devonian seas were ruled by another class of primitive piscine vertebrates, the Placodermi. These also possessed armour plating, but they had a more or less ossified internal skeleton, an upper and a lower jaw and dental plates bearing spikes and hooks. On the head plate there were two eyes and paired nostrils, while on the anterior part of the body there was at least one pair of fin-like limbs. Placoderms do not seem to have had any serious competitors in Devonian seas. They were good swimmers, the majority were predacious, and it is therefore not by chance that their increasing prosperity was accompanied by the decline of many groups of invertebrates, e.g. trilobites, nautiloids, etc. They lived in all the seas, but seem to have preferred shallow inlets and the littoral zone. In the late Devonian they actually colonized brackish and freshwater basins on the continents, in places where 'old red sandstone' formations were developed. Placoderms already appeared as a fully evolved group in upper Silurian seas, but did not yet play an important role. As mentioned above, their peak period was the Devonian, and their last representatives disappeared at the very beginning of the Carboniferous.

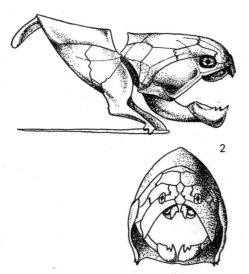

The Devonian genus *Bothryolepis,* known from most parts of the world, has the typical body structure of members of the order Antiarchi. The bony armour on the head and the anterior part of the thorax were fused, as in *Bothryolepis canadensis* (1), making the head immovable. The posterior part of the body was covered with fine scales. The paired, fin-like pectoral appendages were composed of a large number of small plates.

The genus *Dinichthys (Dunkleosteus* — 2) represents the order Arthodira. Its members, over 5 m long, have been found in upper Devonian strata in Europe, Asia and North America. In this group the head and body armour articulated. The pectoral fins consisted of a single segment. The jaws were equipped with spiked dental plates (not yet with teeth), showing that these enormous creatures were predacious.

In lower Devonian seas there are small types of placoderms measuring only about 30 cm, with a flat-sided body and wide fins, and resembling present-day rays. One such type is *Radotina kosorensis,* parts of whose scaly armour (3) have been found in lower Devonian limestones in Bohemia.

Xenacanthus bohemicus

Chordata
Acanthodii — Elasmobranchii

Towards the end of the early Palaeozoic, the first, more highly organized fishes began to appear among primitive fish-like vertebrates, and in the late Palaeozoic they spread throughout the seas and invaded freshwater rivers and lakes. They already had a strong cartilaginous internal skeleton, ossified in places, and their skin was covered with minute, rough or spiny placoid (plate-like) scales which evidently developed in connection with reduction of the outer armour. Another important factor was the formation of true teeth. The most primitive of these fishes were the members of the subclass Acanthodii, which were usually only a few centimetres long, although some may have measured as much as 2 m. In front of every fin grew a thick, conspicuous dermal spine. The eyes, which were situated on the sides of the skull, were framed by a ring of bony elements known as a sclerotic ring. The remains of acanthodians are found chiefly in Devonian sediments, but the last members of the group survived up to the end of the Palaeozoic.

1

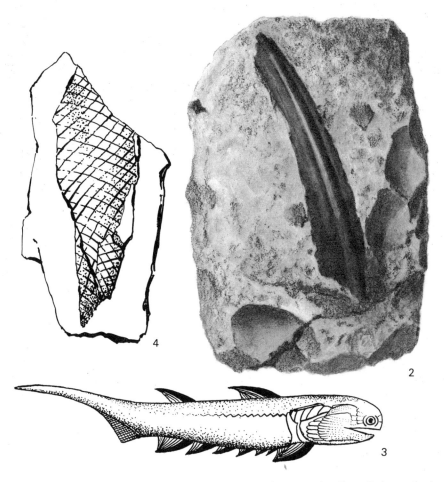

The members of the genus *Xenacanthus* were relatively large elasmobranchid inhabitants of Permo-Carboniferous rivers and lakes. Parts of their skeletons and typical sharp, two- or three-pointed teeth are found in freshwater sediments in almost every part of the world. The index species is *Xenacanthus bohemicus* (1), with a cylindrical body over 1 m long and a large skull with a posteriorly directed, movable spine.

Isolated, curved fin spines up to 30 cm long, which may have belonged to large acanthodians, are frequently found in Devonian marine limestones in Europe, North America and north Africa. They were originally described from Bohemia as *Machaeracanthus* (now *Orthacanthus*) *bohemicus* (2).

Climatius (3) was a typical small acanthodian, only about 10 cm long. It is remarkable for its numerous paired fins reinforced with wide, curved dermal spines. Many species are already known from the Silurian period, and they are common in Devonian strata in Europe, North America and Spitzbergen.

From Permian and Carboniferous freshwater deposits in central Bohemia we know only the skeletal remains of primitive sharks, but their egg capsules, described as *Paleoxyris* (4) have been found. The latter are spindle-shaped and bear characteristic spiral ridges.

Paramblypterus rohani

Chordata
Osteichthyes

Representatives of the true bony fishes (Osteichthyes) also lived in Palaeozoic waters, but it was not until subsequent eras that they really began to flourish. Among Palaeozoic forms, special mention should be made of the extinct order Palaeonisciformes, which played an important role in their day. They were freshwater fishes with an incompletely ossified skeleton and a fusiform, relatively high body covered with smooth, strong rhomboid scales. The scales were made of ganoidin, and the fish were named ganoids after them. The paired fins of palaeoniscoids structurally resembled the fins of present-day fishes, but their characteristic feature was their asymmetrical (heterocercal), deeply concave caudal fin. The oldest and most primitive palaeoniscoids appeared in middle Devonian freshwater sediments, but they attained maximum distribution in Carboniferous and Permian rivers and lakes, especially in the northern hemisphere, where they were then the dominant element. From the beginning of the Mesozoic era the whole order started to decline, and its members died out in the lower Cretaceous. Their specialized or degenerate descendants still live in the waters of tropical Africa; they broke away from the main phylogenetic branch some time during the Permian period and they include the sail-finned *Polypterus bichir* of the Nile and the related, eel-like *Calamoichthys*.

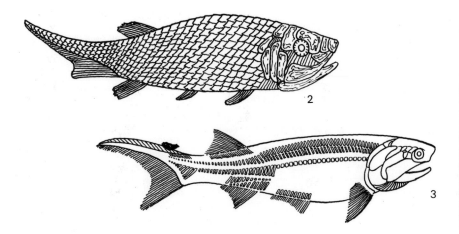

1

In Carboniferous-Permian shales accompanying coal seams in central Bohemian coalfields we sometimes find many whole specimens of *Paramblypterus rohani* (1) measuring 15—30 cm. Related species are known from strata of the same age throughout the whole of Europe.

Another distinctive species of European upper Carboniferous formations is the tiny *Pyritocephalus sculptus* (2), only 10 cm long, remarkable for the fine, concentric, but undulating grooves on the bones of its head.

The diagram of a primitive palaeoniscoid of the genus *Cheirolepis* from Devonian strata in Europe and America shows the already relatively complicated structure of these fishes' internal skeleton (3). Species of this genus are fairly common in freshwater slates and sandstones; they have a long, fusiform body about 30 cm long, small, but numerous teeth and comparatively strong jaws and are covered with fine ganoid scales.

181

Ctenodus obliquus

Chordata
Osteichthyes

The swamps round tropical rivers in Australia, South America and Africa are still inhabited by curious fishes, up to 175 cm long, which do not mind living in muddy water with a poor oxygen supply, as they possess a pulmonary sac as well as gills and all they need to do is to surface and take in atmospheric oxygen. They do not even care if their pool dries up during the dry season, but simply ensconce themselves in a mucus-lined cylindrical capsule and wait for conditions to improve. When the rains come and the pool fills up again, the water dissolves the capsule and the fish 'come back to life'. Natives dig up the fish and eat their tasty flesh during the dry season, and, although it may seem a strange to go fishing with a hoe, it is the most effective way, since lung fishes (order Dipnoi, which means two ways of breathing) are otherwise very quick and predacious. The three genera still extant are only the poor remains of a once very large order. They are indeed living fossils; for instance, the Australian lung fish *Neoceratodus* was known only from fossils long before the first living specimens were found. The oldest members of the order Dipnoi come from the Devonian period, when they colonized freshwater habitats all over the world. Their evolution progressed in the Carboniferous period, but declined somewhat in the Permian. At the beginning of the Mesozoic era they enjoyed a new period of prosperity, but their subsequent decline has continued ever since.

Probable lung fish remains are found chiefly in Carboniferous-Permian sediments originally deposited in swamp environments. The majority are head bones, fragments of ribs and the especially characteristic large, plate-like teeth with 4—6 pectinate ridges. Fig. 1 shows a fragment of cannel (a type of coal) containing jawbones, complete with

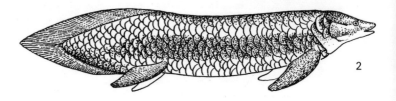

2

1

3

teeth, and fragments of head bones of *Ctenodus obliquus,* a widely distributed freshwater species known in Europe, North America and Australia. The long paired fins were powerfully developed and allowed the fish to crawl in the mud as well as to swim. The diagram 2 shows the typical body shape and organization of the living lung fish of the genus *Neoceratodus* from Australia.

The irregular part of the bony cephalon of *Gompholepis panderi* (3) was once thought to be a fossilized fragment of upper Silurian-Devonian Dipnoi. Today the remains are considered to be fragments of the armour of members of the class Agnatha.

Latimeria chalumnae

Chordata
Osteichthyes

In 1938, when the captain of a fishing schooner off the coast of Madagascar gave orders to lower the fishing nets, he did not dream that this would make his name in palaeontological history, even when part of the haul was a strange fish about 1.5 m long, with large eyes and a steely blue body covered with large, strong scales, which was brought up from a depth of some 60 m. What did strike him was that the unknown fish was unusually active, moved on its stout, stumpy fins and kept trying to scramble back into the sea. Captain Goosen was a sensible man, however, and the fish did not land either back in the sea or in the market, but in the highly qualified hands of Professor J.L.B. Smith, who was amazed to find that it was a coelacanth, a member of the order Crossopterygii, which according to palaeontological evidence already lived during the Palaeozoic era and apparently died out in the Mesozoic era, some 70 million years ago. And now, all of a sudden, here was this 'living fossil'. Crossopterygians were primitive fishes which illustrate how vertebrates, during their evolution, may have moved from the water to the dry land. Like Dipnoi, they could breathe atmospheric oxygen, and their paired fins were so thick and strong that their owners were possibly able to migrate 'on foot' from drying pools to better sources of water. Most crossopterygians were predators; their body was covered with tough cosmoid scales and their skull with bony armour-plating. They first appear in the Devonian, but their full development does not start until the Permian and continues into the Mesozoic era. No fossilized remains are known from the Cretaceous period onward, but we are now familiar with the famous, present-day representative.

The extant crossopterygian, described as *Latimeria chalumnae* (1), gives us a good idea of what fossil types looked like. Today we know several dozen specimens of *Latimeria* from the region of Madagascar and the Comoro Islands, and they are still being subjected to a thorough study.

The diagram of the upper Devonian genus *Eusthenopteron* (2) shows the basic organization of the body of primitive Palaeozoic crossopterygians. The most noteworthy features are the ossified spine and the marked fin bones on which the fins themselves were set and which the dorsal fins also possessed. The remains of several species of these fishes, which measured about 1.5 m, have been found in Europe and in North America.

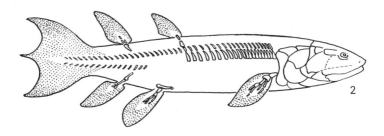

Branchiosaurus salamandroides

Chordata
Amphibia

Amphibians were actually the first vertebrates which finally succeeded in invading the dry land — not completely, however, as they still lay their eggs in the water and their larvae develop there. Although adult individuals live on the land, they remain close to water and in a damp habitat. They stand on a very low rung of the evolutionary ladder, and their oldest representatives in particular, which are known from Devonian formations, still have many features of their piscine forebears. The chief, late Palaeozoic, primitive amphibians are the members of the extinct subclass Labyrinthodontia, which lived in Carboniferous-Permian swamps and forests. They are known as 'stegocephalians' (meaning roof-skulled), because the top of their head was encased in bony armour. The majority measured only 15—20 cm, although species over 2 m long are also known. Their body was lizard- or salamander-like, with a wide head and a long tail. In addition to large paired eyes framed in a sclerotic ring, there was a small, single parietal eye on the head. The body was covered with armour formed of variously shaped scales. The fore limbs had four toes and the hind limbs five (making the five-toed vertebrate limb at least 300 million years old). The small, but sharp, conical teeth show that labyrinthodonts were predacious. After a time of great prosperity in the Carboniferous-Permian period, the entire subclass died out at the end of the Triassic period, i.e. at the beginning of the Mesozoic era.

In European upper Carboniferous and Permian strata are quite often found complete skeletons of small salamander-like labyrinthodonts belonging to the genus *Branchiosaurus*. A number of species have been described in central Bohemian bituminous coal shales, the commonest being *Branchiosaurus salamandroides* (1), about 5 cm long. Today we know that this 'genus' actually comprises several genera of these primitive 'stegocephalians'.

The larval stages of several primitive, not properly identified labyrinthodonts (2) found in Carboniferous-Permian sediments have also been described as *'Branchiosaurus'*. It is interesting that these larvae still have outer dendritic gill arches, which were lost in adulthood in just the same way as they are in some amphibians today.

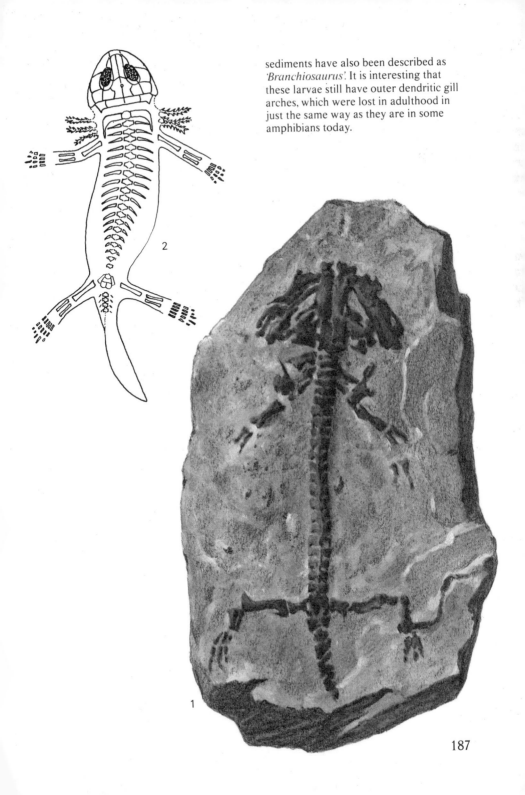

Anthracosaurus dyscriton

Chordata
Amphibia

The order Anthracosauria was a specialized group of Carboniferous-Permian labyrinthodont amphibians. It comprised relatively large forms somewhat like crocodiles in appearance and often up to 2 m long. Anthracosaurians had a long, tapering head, small eyes, large paired nostrils, strong teeth with characteristic deep folds in their enamel and a skeleton with vertebrae that completely ossified in adult life. They were thus a higher stage in the evolution of labyrinthodont amphibians, and, in the opinion of some research workers, they were the representatives of the evolutionary line from which the reptiles branched off. Judging by the structure of their slim, lizard-like body and their strong teeth, most anthracosaurians seemed to have been relatively quick, nimble and rapacious animals and a menace to other, smaller types of 'stegocephalians'. The oldest anthracosaurian amphibians already appear, though seldom, in European early Carboniferous strata. In the late Carboniferous the entire group suffered an upheaval on both the European and the North American continent. As regards their mode of life, anthracosaurians seem to have been restricted to late Carboniferous swampy lowland forests. In the Permian period, when the climate became drier and the forests disappeared, the specialized anthracosaurians died out with them.

In European and North American late Carboniferous strata we often find quite large numbers of fragments of the skull, and occasionally parts of the skeleton, of the typical genus *Anthracosaurus*, which was characterized by a tapering triangular skull and a body over 1 m long. One of the geologically youngest species was *Anthracosaurus dyscriton* (1), whose skull, measuring over 20 cm, was found in early Permian coal slates in northern Bohemia.

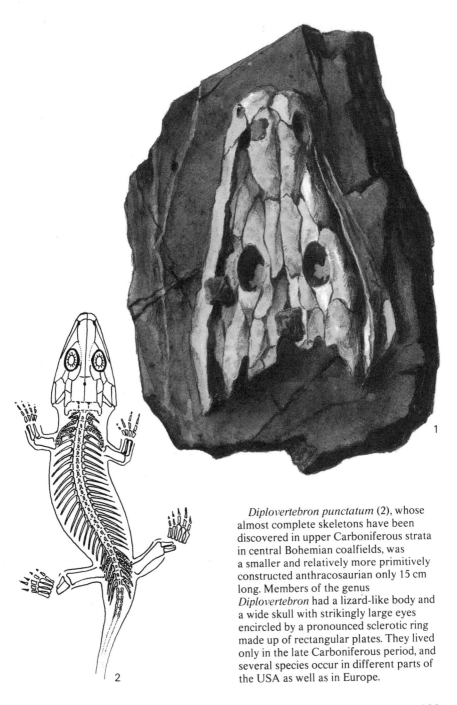

Diplovertebron punctatum (2), whose almost complete skeletons have been discovered in upper Carboniferous strata in central Bohemian coalfields, was a smaller and relatively more primitively constructed anthracosaurian only 15 cm long. Members of the genus *Diplovertebron* had a lizard-like body and a wide skull with strikingly large eyes encircled by a pronounced sclerotic ring made up of rectangular plates. They lived only in the late Carboniferous period, and several species occur in different parts of the USA as well as in Europe.

Discosauriscus potamites

Chordata
Amphibia

Finds made during palaeontological research on the flora of lower Permian strata in Texas in the USA included the skeletal remains of amphibians about 75 cm long, which, as a group, proved very remarkable because the structure of their skeletons combined primitive amphibian characters with explicitly reptilian features. They had 'stegocephalian' skulls with a bone-covered crown, a single parietal eye (parietal foramen), labyrinthodont teeth with characteristic, wrinkled enamel and scaly body armour, etc. Their main reptilian feature was the single articular process on the occipital bone (as against two in stegocephalians), which allowed much greater movement of the head. These details all show that *Seymouria,* as the find was called, was a higher type of amphibian close to the reptiles. Today we know many such genera, from Europe and Asia as well as from North America. They are assembled in the separate suborder Seymouriamorpha, whose incidence is virtually restricted to the Permian period; a few, rare early representatives have been found in uppermost Carboniferous strata. Seymouriamorph amphibians do not seem to have had a direct need of water, and adult individuals were probably able to live in dry regions.

Early Permian bituminous limestones in Moravia have already furnished several thousand remains of small seymouriamorph amphibians of the genus *Discosauriscus,* which are only some 20—30 cm long. The almost complete skeleton belongs to the species *Discosauriscus potamites* (1), which had a relatively narrow skull with large eyes and a slim, lizard-like body covered with small, rounded scales.

The original reconstruction of the most abundant Moravian species, *Discosauriscus pulcherrimus* (2), is based on numerous actual finds. Unlike the preceding species, it had a round-tipped triangular head and a more salamander-like body. The adults of both species lived on dry land and hunted their food in the undergrowth of the early Permian flora beside rivers and lakes.

The reconstruction of the skull of the typical genus *Seymouria* from Permian strata in Texas shows the still generally 'stegocephalian' character of the bone organization (3).

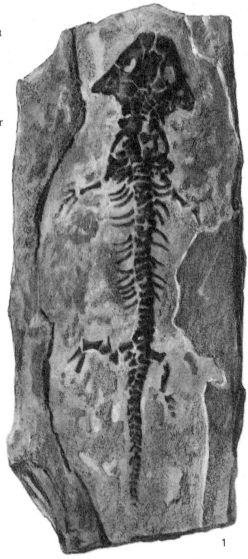

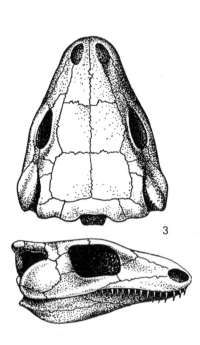

Urocordylus scalaris

Chordata
Amphibia

The subclass Lepospondyli is an independent evolutionary branch of stegocephalian amphibians which broke away from its ancestors (probably crossopterygian fishes) at some point early in the Devonian period. These amphibians had a relatively well ossified skeleton; the spine was composed of spool-shaped vertebrae which constricted the spinal cord. The many species in this group are often curiously constructed; alongside lizard- and salamander-like forms, they also include snake-like forms in which the scapular belt, as well as the limbs, was completely reduced. They generally did not measure more than 10—20 cm and were seldom longer than 0.5 m; they also include the smallest amphibians ever known, which when adult measured less than 2 cm. Most of these stegocephalians seem to have been restricted to swampy lowland 'anthracite' forests of horsetails, club mosses and primitive ferns. Although remains of their oldest representatives have been found in lower Carboniferous strata, their peak period was the late Carboniferous. The advent of the late Carboniferous forests was accompanied by a tremendous development in lepospondylous amphibians. At the beginning of the Permian period, when the climate grew rapidly drier and the swampy forests receded, these interesting animals died out.

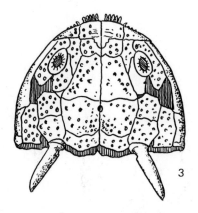

3

Urocordylus scalaris (1), with its wide, triangular head, slender body and long, wide tail, was evidently equally active in the water and on the land. The numerous remains found in upper Carboniferous coal shales in central and western Europe have permitted good reconstruction of this species.

The interestingly constructed lepospondylous amphibian *Keraterpeton* had two long, sharp spines on the back of its skull. The many species of this genus were fairly widespread in the Carboniferous of Europe and North America.

Part of a skull of the typical species *Keraterpeton crassus* (2), with pronounced occipital spines, was found in upper Carboniferous coal shales in Central Bohemia (Fig. 3 shows the reconstructed skull).

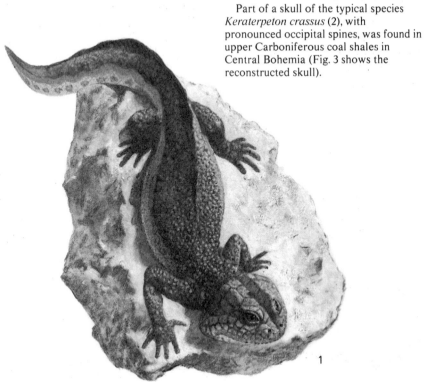

Dolichosoma longissimum

Chordata
Amphibia

The order Aistopoda was a peculiar, specialized group of Carboniferous lepospondylous amphibians with a slender, serpentine body over 75 cm long and with rudimentary or no limbs. Like the skulls of other 'stegocephalians', the top of their triangular skull was enclosed in bony plates and their spine was composed of up to 200 vertebrae to which thin and partly rudimentary ribs were attached. Only the dorsal side of the body was covered with small scales, i. e. there was no ventral armour. The teeth were simple, small and conical or slightly curved. Aistopods did not play a marked role in Carboniferous animal associations, although in places their remains are fairly common. To date we know of three genera, which lived only during the Carboniferous and the beginning of the Permian period in Europe and North America, but the oldest members of the group date back to the early Carboniferous, so the order as a whole can be ranked among the oldest amphibians known. The difficult question is how these amphibians lived. They were probably predacious. It was once supposed that they were completely aquatic and lived chiefly on fish, but the majority of present-day vertebrate palaeontologists are of the opinion that adult individuals crawled and hunted on the marshy ground in Carboniferous swamps.

Almost complete skeletons over 50 cm long, belonging to delicately built individuals of the species *Dolichosoma longissimum* (1 — a skull), were found in upper Carboniferous coal shales and cannel coal horizons in central Bohemian coalfields; the finds allowed the reconstruction of these remarkable amphibians (2).

Figure 3 shows the reconstructed skeleton of a very similar genus which lived in the Carboniferous of North America as well as in Europe. It was named *Ophiderpeton* and differed from the preceding species in respect of its fewer vertebrae (it had only about 100) and the lateral processes on its ribs.

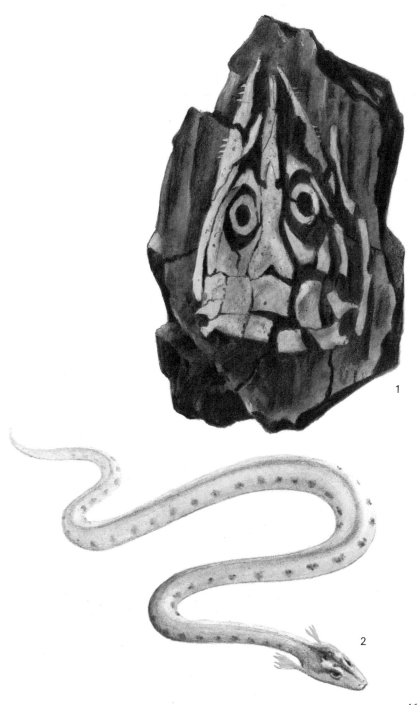

Edaphosaurus sp.

Chordata
Reptilia

The highest level in the evolution of animals during the Palaeozoic era was attained by the reptiles (Reptilia). No traces of them have yet been found in the early part of the era, but in the Carboniferous-Permian periods they already numbered many genera and species and often produced weird or monstrous forms. Reptiles did not directly need the presence of water, as did amphibians, and were fully adapted to a terrestrial existence. In addition to primitive types still strongly reminiscent of 'stegocephalian' amphibians in particular, Carboniferous-Permian forests, river banks and lakesides and higher and drier places were inhabited by reptiles with the most diverse appearances and with a more modern body structure already resembling that of mammals. These were pelycosaurs (order Pelycosauria), a closed group which appeared at the end of the Carboniferous period and died out in the middle Permian. Its members were mostly only about 1 m long, but species measuring over 3 m are also known. In appearance and habits, some of the more lightly built pelycosaurs resembled the predacious present-day monitors *(Varanus)*. Massive, cumbersome and often very strange-looking forms were generally herbivorous. The greatest numbers of pelycosaurian reptiles are known from the USA; lower Permian red sandstones in Texas and New Mexico in particular have furnished a quantity of skeletal remains of different species. Other remains have been discovered in Europe and South Africa.

Edaphosaurus was one of the monstrous pelycosaurian genera in lower Permian forests. These herbivorous reptiles, some of which were over 3 m long, had extremely long vertebral spines which formed a kind of high, bony rampart, covered and joined by skin, down the middle of their back. The genus *Edaphosaurus* has been found in Europe as well as in North America. One of the

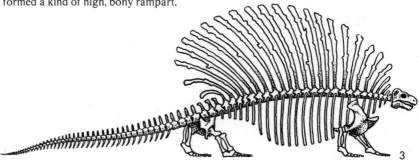

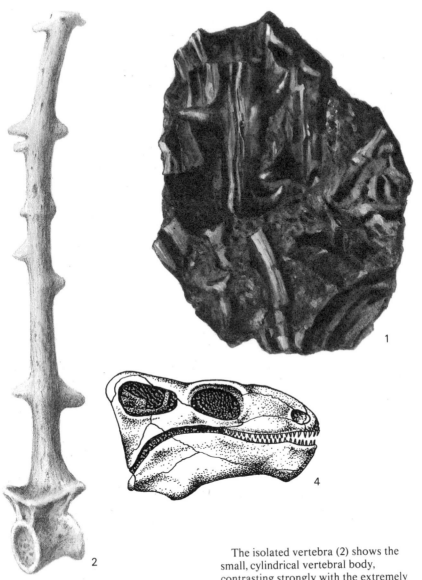

rare European finds — parts of vertebral spines with the lateral processes typical of this genus — came from anthracite coal seams in Moravia (1).

The isolated vertebra (2) shows the small, cylindrical vertebral body, contrasting strongly with the extremely high vertebral spine.

The diagrammatic reconstruction of the skull and whole skeleton of *Edaphosaurus* shows the general organization of the body and the appearance of these curious reptiles (3, 4).

Lycosuchus sp.

Chordata
Reptilia

A large part of southern Africa is composed of Permo-Triassic strata comprising sandstones, conglomerates and slates hundreds of metres deep, known in geological literature as the Karroo formation. It has furnished valuable remains of bygone terrestrial, and in particular Permian, vertebrates, including skulls and parts of skeletons which aroused considerable interest among experts. In these strange animals, the skull already had a completely closed bony base, and in predacious forms the teeth were differentiated into incisors, canines and molars, i. e. explicitly mammalian characters. A long discussion followed as to whether these were primitive, late Palaeozoic mammals, but a thorough study showed that the fossil remains were of reptilian origin. Their owners, which were called Therapsida (order), broke away from pelycosaurs some time during the Permian period and formed an evolutionary branch which only resembled mammals as far as their organization was concerned. It was as if nature, in these mammal-like reptiles, was trying out one of the ways to a new type of animal. It turned out to be a blind alley, however, and therapsids died out at the beginning of the Jurassic period, i. e. in the middle of the Mesozoic era, without leaving any descendants. At the time of their greatest prosperity, i. e. in the Permian period, the therapsids were a vital group; they formed a number of species in Europe (especially in the region of the Urals in the USSR) as well as in southern and eastern Africa, and we even know of Triassic specimens from South America.

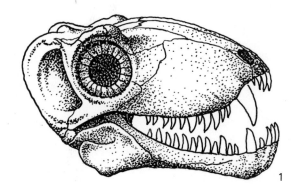

1

2

The predacious, middle Permian therapsid *Phthinosuchus* from the USSR was about 1.5 m long and its skull (1), which measured about 20 cm, still displayed primitive reptilian characters, such as the orbital sclerotic rings. On the other hand, the narrow rostral part of the skull and the development of a distinct snout, together with the differentiated dentition, are mammal-like features.

One of the commonest representatives of therapsid reptiles in middle Permian strata in southern Africa was the somewhat larger, rapacious and lightly built *Lycosuchus* (2), which measured over 2 m. The tapering triangular skull is notable for the functionally differentiated teeth and particularly for the sharp canines, which were up to 5 cm long.

Parasitism and injury

Modern palaeontology long ago stopped studying the animals of ancient seas and continents in isolation, simply on the basis of descriptions of the fossil remains of individual species. Scientists now try to reconstruct primaeval life in all its rich diversity and to investigate the development and the reciprocal relationships of animal associations and individual organisms. Here again, the careful collection and study of fossil remains can furnish a great deal of interesting and important information. For instance, on the remains of the oldest Palaeozoic organisms we already find various pathological tumours and swellings on the outer or inner surface of shells — deformities indicative of the activity of parasites. A comparison with analogous conditions among recent organisms will even give us some idea of the actual parasites themselves. In other cases we find shells or parts of shells or other skeletal material which are deformed, broken off or damaged in some other way indicative of mechanical damage or of injury by a predator; they are injuries which occurred while the animal was still alive and which generally healed but left traces on the body casing.

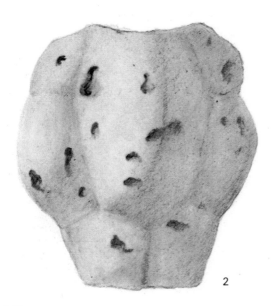

2

On the stalks, and occasionally on the cup plates, of central Bohemian Devonian crinoids there are large numbers of pathological swellings caused by organisms which bored into the crinoid's skeleton and lived as parasites inside the body (1). These frequently bulging or villoid swellings sprout from the stalk and completely deform it. It has not been resolved definitively which organisms were responsible, but it is generally presumed that they were parasitic worms belonging to the Myzostomatida group.

Not even the tiny crinoids of the species *Junocrinus globulus* (2), from the Bohemian Devonian family Ramacrinidae, were spared. The enlarged cup (which is actually only 3 mm across) shows a number of small, oval depressions produced by ectoparasites (i.e. living on the exterior of the host), which attached themselves to the plates of the cup and destroyed its calcareous skeleton. Similar depressions caused by parasitic fungi of the genus *Vioa* are known on the body cases of present-day marine animals.

The relief of the caudal plate of the Silurian trilobite *Scutellum* is an example of mechanical injury. The wound has healed, but the course of the pleurae and pleural furrows at the site of the injury has been affected, and the pygidium itself is slightly asymmetrical (3).

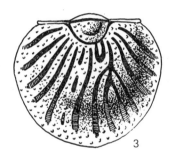

Commensalism and symbiosis

Commensalism and symbiosis are relationships between organisms of which fossils provide plenty of evidence. In this case the host organism is not attacked or directly endangered, as in the preceding chapter, but the guest-host relationship is almost equally balanced and can actually be described as a kind of cooperation. Commensalism is an intimate relationship, but without much mutual influence, between two organisms, which is of advantage to one (the commensal), but does not harm the other. It can be demonstrated already in Palaeozoic associations and is also common in nature today. All organisms which live attached to the hard body case of other organisms can be regarded as commensals.

Symbiosis is an intimate relationship between dissimilar organisms beneficial to both organisms involved. Mutualism is an extreme form of symbiosis in which the guest and the host have become so completely dependent upon each other that neither can survive under natural conditions without the other. The knobbly, calcareous alga *Sphaerocodium* from Scandinavian Silurian strata, which is actually a complex of several organisms, is a fossil example of mutualism. Far more examples are known among recent animals, e.g. among insects.

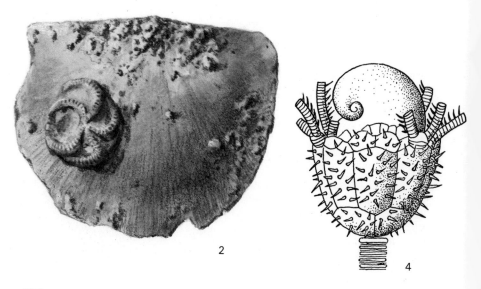

2

4

Fig. 1 shows a typical case of commensalism in fossils from Devonian marine strata in the USA. A colony of small, chain-forming bryozoans of the species *Intrapora irregularis* is growing on the firm base provided by a large, flat clump of corals of the genus *Hernodia*.

The small edrioasteroid echinoderms of the middle Ordovician species *Carneyella pilea* from the USA, which have attached themselves to the large, wide shells of a brachiopod belonging to

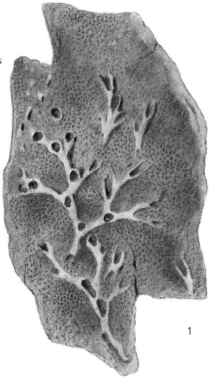

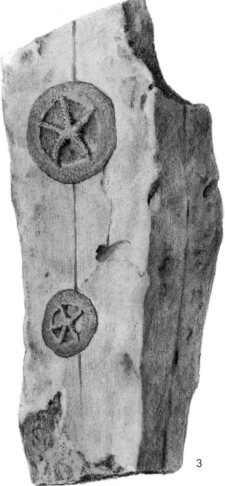

the genus *Rhipidomena* (2), are another example of commensalism among sessile animals.

Originally, sessile commensals sometimes settled on the body cases of swimming or floating hosts. The cases of conularians from Bohemian Ordovician strata often carry several specimens of sessile, discoid edrioasteroid echinoderms (3).

The interrelationship between upper Silurian and Devonian platycerate gastropods and certain crinoids can be classified as symbiosis. The gastropods lived firmly attached to the roof (tegmen) of the cups of crinoids (e.g. of the genus *Arthroacantha*) (4) and lived on their excrement, thereby removing it and doing the host a good service.

Traces of the activities of fossil organisms
Exogenic traces

As well as examining the remains of actual extinct animals, that is to say, their shells and skeletons, palaeontologists have also begun to investigate traces of their activities, i.e. manifestations of their habits, such as tracks, grooves made by crawling, the passages of burrowing animals, resting trails, signs of feeding habits, etc. These phenomena are grouped together under the collective term trace fossils, which are in principle divided into exogenic traces — surface traces — and endogenic — traces within the sediment. It is very rare to find such traces together with the animal that made them, and they are therefore generally evaluated according to their resemblance to the tracks of extant organisms, or on the basis of a morphological analysis. It is becoming increasingly clear that the study of trace fossils is of great significance for palaeontology, chiefly because it supplements the picture of life in bygone geological ages or in types of environments in which, for one reason or another, no other remains of fossil organisms have been preserved. Trace fossils occur in practically every type of stratum and are known from the earliest geological times. At present, however, their detailed classification is difficult and is subject to intensive study.

A rare find of surface trails, together with the trilobites responsible for them, was made in lower Cambrian slates in Pakistan. The trails were described as *Dimorphichnus obliquus* (1), but the trilobites that made them could not be identified because of their poor state of preservation.

Remarkable stellate trace fossils found in Bohemian Ordovician strata were described as *Asteriacites fallax* (2). They are actually impressions made by the bodies of asteroids or brittle stars resting on the soft sediment. Similar trace fossils have been found in strata of the same age and in younger strata till Tertiary throughout Europe and in North America.

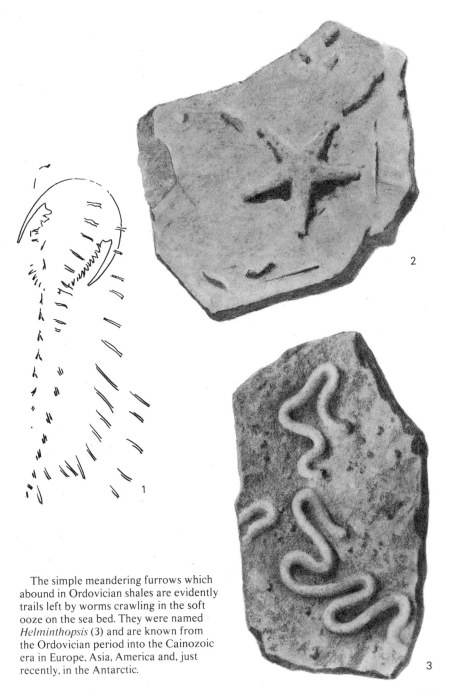

The simple meandering furrows which abound in Ordovician shales are evidently trails left by worms crawling in the soft ooze on the sea bed. They were named *Helminthopsis* (3) and are known from the Ordovician period into the Cainozoic era in Europe, Asia, America and, just recently, in the Antarctic.

Traces of the activities of fossil organisms
Endogenic traces

A large proportion of marine invertebrate animals live directly in the sediment on the sea bed, either in the soft ooze or in sand near the shore. In this respect fossil organisms were no different, and, consequently, today we find sandstones and shales with traces of a whole system of vertical, oblique or horizontal shafts and tunnels. These inner trails or endogenic traces were usually the burrows of various worms, for example, or were made by organisms which lived on organic substances in the ooze and literally ate their way through the sediment. Quite often the sediment was re-distributed by their activity, so the study of these trails is also important for sedimentology. Endogenic traces appear in sediments from the beginning of the Palaeozoic era, i.e. in lower Cambrian strata, and some are known from Proterozoic sediments. Tunnels used to be grouped together under the term 'chondrites' and vertical shafts and resting trails under the term 'scolites'. Today, several hundred types have already been differentiated and are being studied all over the world. In some places they actually form distinctive horizons which can also be used for local stratigraphic comparison of the various layers.

1

2

3

Long vertical shafts about 5 mm wide, sometimes still showing rows of transverse rings, are often found in European, Asian and North American Palaeozoic quartzites. They were described as *Tigillites* and are evidently the resting trails left by sandworms in seashore sediments. *Tigillites vertebralis* (1) is a characteristic trace fossil of Bohemian Ordovician quartzites.

The diagram shows another type of tubular resting trail with a U bend and a dumbbell-shaped opening. Casts of these trails have been found in Bohemian middle Ordovician silt sediments and were described under the collective name *Rhizocorallium* (2). Identical shelters are built in argillaceous-sandy parts of the sea bed by extant marine worms of the genus *Polydora* (3), so that their form has remained unchanged for 480 million years.

FOSSILS TELL A STORY

Although they are mute it is no exaggeration to say that fossils 'speak', since they provide a wealth of information which up to quite recently palaeontologists would never have dreamed possible. The view that the palaeontologist merely looks for and collects quaint traces and examples of life in past geological ages, describes them, classifies them in broadly related groups, i.e. higher systematic units, and then simply pores over them with delight or silent wonder, went out about 100 years ago, although it still crops up from time to time.

Fossils are undoubtedly of great practical significance for geological research. There are several ways of dividing geological time, but biostratigraphy (the division and determination of the relative stratal age from the palaeontological contents of a stratum) is still the most exact and the most reliable; and the fact that it is the cheapest is not to be despised either. Biostratigraphic data are widely used both for basic geological mapping and in the search for deposits of economically important raw materials such as oil, coal, salts, phosphates and sedimentary uranium and iron ores, and also in hydrogeology, etc.

Good biostratigraphic data can help to clarify the tectonics of a region, e.g. fractures and folds in the strata, when strata which once rested on top of each other shifted over each other, overturned or subsided, or, conversely, were thrown up and partly carried away. Even the age of certain parts of transformed, metamorphosed strata can be estimated on the basis of fossil remains.

One of the oldest, firmly upheld geological precepts was that all intensively metamorphosed rocks such as gneiss and mica-schist were primary rocks which dated back to the time when the Earth's crust was first formed. It was accepted up to the time when Scandinavian mica-schist complexes were found to contain the indubitable remains of Palaeozoic trilobites just a mere 450 million years old and even more recent fossils were found in Alpine metamorphic rocks. The 'unshakable' theory thus came to a sudden end, and these finds threw a completely new light on metamorphism and on the age or periodicity of these processes.

Life in the past was just as manifold and dependent upon environmental conditions as it is today. Modern palaeontology does not work solely, therefore, with so-called index fossils, that is to say, with those completely typical of given strata. Where the nature of the material permits, it uses biological methods based on comprehensive evaluations of associated fossils, including computer-based statistical analyses.

I have already mentioned that fossils are a source of biological

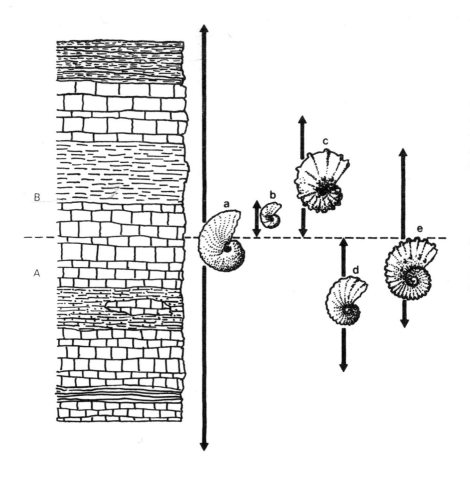

Fig. 2 — an ideal example of the use of fossils in biostratigraphy. The best fossil for the most exact determination of the dividing line between the two formations A and B is type *b*, which has the smallest time range and occurs only in the basal layers of formation B. Types *c* and *d* are also good index fossils, whereas type *e* is suitable only for approximate boundary determination. Type *a* has a very wide stratigraphic range, occurring in both formations, and is therefore relatively unimportant for detailed biostratigraphic work.

information. Our extant fauna and flora are the outcome of well over a million years' evolution, and they cannot be comprehended only in the plane of the present. Palaeontology adds many more planes of time, and it can be generally claimed that from looking at nature three-dimensionally we are beginning to look at it in four dimensions.

The degree of interrelationship of extant organisms can be studied by recent biology (neontology), by deductive methods based on comparative anatomy, functional morphology, ontogenesis (the development of the individual), biochemistry, genetics and other branches of science. Sometimes this leads to different results and hypotheses, which can be verified only by direct comparison with palaeontological facts.

Fossils also furnish important data on the environments in past geological ages, i.e. palaeo-ecological data. In the case of aquatic animals, they not only indicate whether the sediments are of marine, brackish or freshwater origin, but they may also provide detailed information on the degree of salinity of the sea, on its relative depth and temperature, on the direction and strength of the currents and on the oxygen supply, etc.

Some groups of marine animals, such as echinoderms, are particularly sensitive indicators of environmental conditions and so are a good source of information. Terrestrial animals are an excellent guide to the temperature, humidity (or aridity) of their surroundings, climatic changes and seasonal changes in the weather.

This brings us to the contribution of fossils to palaeogeography. A study of animals through the geological ages makes it possible to estimate and reconstruct the geographical image of past landscapes or the distribution of land and sea. These studies are particularly significant today in relation to worldwide research on the character and evolution of the Earth's crust, the movement of continents and of the ocean bed, magnetic pole shifts and other geophysical and geographical problems.

There is also a great future in store for palaeontology in association with lithology (the science of rocks) and geochemistry. The activities of organisms had (and still have) a strong influence on geochemical processes in the air, on the land and in the water. For instance, free oxygen — the first essential of life for higher organisms, including man — is largely one of their products. In addition, organisms played (and play) an important role in the geochemical cycles of hydrocarbons, sulphur, phosphorus, carbon and other elements. Organisms supply the material, or at least the stimulus, for the formation of various important sediments and predetermine their structure, so that a thorough study of sedimentary rocks, and especially industrially important carbonates, is impossible without taking into account the role of organisms.

There is a great deal more that could be said about the role of fossils and palaeontology in worldwide research on life and the environment, particularly as regards their future role in the protection and preservation of the environment, with special reference to the survival of man as a species. These questions are dealt with in detail by specialized palaeontological publications.

This book has a somewhat different purpose. It primarily concerns the fauna of the Palaeozoic (Primary) era and only occasionally digresses into the Mesozoic (Secondary). The Palaeozoic is the oldest era in the Earth's history which is well documented by fossils. In fact, its beginning also marks the beginning of explosive development of animal life on our planet. Although remains of some groups of invertebrate animals are known from older strata, i.e. from Proterozoic times, they are so different from Palaeozoic animals and are relatively so few in number that it is necessary to take the beginning of the Palaeozoic as a kind of dividing line in the general history of life on the Earth. How it came about that practically all invertebrate phyla, and apparently some primitive chordates suddenly appeared in the early Cambrian period is something that has puzzled scientists ever since palaeontology and biostratigraphy became sciences.

It would be a mammoth task to describe all the hundreds of thousands of species of aquatic and terrestrial Palaeozoic animals. In any case, there is no one author who could do it. For instance, the many volumes of standard compendiums like the Treatise on Invertebrate Palaeontology (which are of only classification and identification value) are the work of teams of experts from all over the world, who are continuously supplementing and revising them as palaeontological research increases in depth and extent. The aims of the present publication are purely to give the reader a glimpse of the beauty and diversity of the faunas of ages separated from the present by 570 to 220 million years, and to stimulate his imagination a little, so that from those incomplete and often uninteresting-looking fossilized remains he can conjure up the splendid abundance of the creatures that inhabited the ancient seas and continents.

A FEW WORDS ON HOW TO COLLECT FOSSILS

There is an old European proverb that says that if you want a bear's skin, you must first catch the bear. Similarly, if you want a collection of palaeontological material, you must first collect the fossils. In the latter case there is the advantage that the fossils are dead and cannot run away, but even so, each day missed is a day lost. Every day new parts of quarries are being opened up, ground is being excavated, e.g. for the foundations of buildings, and horizons rich in fossils are being laid bare in places to which we should otherwise never have access, but the time during which they are available is all too short. Classic sites, on the other hand, age and deteriorate and gradually lose their importance. The quality of palaeontological research results is therefore directly dependent on possibilities of collecting fossils and, of course, on the methods by which they are evaluated.

Fossils are collected either for purely scientific purposes, or as a hobby. The aims of scientific, professional and documentational collecting were mentioned in the introduction. In this case the efforts of palaeontologists are directed towards securing comprehensive data of use to biology, stratigraphy, ecology and palaeogeography, etc. Collecting as a hobby takes two forms. In one form, which is fortunately less common, only decorative or rare specimens are collected, so that the collector can show them off to envious fellow fossil collectors. However, such collections are not valueless. If they are properly and carefully catalogued and identified, they may be of considerable scientific value. Unfortunately, that is seldom the case, and the collector is dismayed or indignant when the expert pityingly shakes his head over his treasures.

The other group of amateur collectors actually forms a large team of collaborators who often render invaluable aid to palaeontological research. They pry deep into their subject and take their hobby very seriously. They keep full and accurate records of their finds, especially as regards the locality, which is extremely important. Such collections may frequently contain whole associations, either from known sites or from chance excavations which the expert may never hear about or which he has no time to visit. These amateurs often engage in serious study and quite frequently achieve excellent results, either by their own unaided efforts or by working together with an expert.

To those who are interested in fossils and are attracted by the adventure of probing into the past, we should like to give a little initial advice. Fossil collecting and palaeontological research, like any other type of investigation, naturally requires patience and perseverance, and although the preliminary results may be modest, we should not forget that even the biggest and most famous collections started with item number 1. The serious collector must also have powers of observation. The old method was first to collect the fossils and then to sort them out in the laboratory; but today the reverse system is used, i.e. the terrain is inspected first and the fossils are collected afterwards. In the original records made on the spot, the first thing to note is whether the fossil is loose or in situ within the exposure. The next thing is to determine whether it is in life position (which occurs in the case of sessile and fused animals like corals [Anthozoa], and bryozoans, etc.), or had been transported and secondarily deposited. The original current can sometimes also be determined from the orientation of the fossil. If we collect fossils from horizons of different geological ages, we can determine which layers or horizons contain fossils and which do not. We naturally try to find as many different genera and species as possible from individual fossiliferous horizons and thereby to determine the character of the associations and any vertical (= temporal) or lateral (= spatial) changes which may occur in them. If the various fossiliferous horizons also differ petrographically, i.e. as regards the character and composition of the rocks, such findings provide a good guide in later palaeontological research over a wider area around the original site.

When collecting fossils, the petrographic composition of the fossiliferous horizons must always be taken into account. Some rocks, like slates, silts, shales, etc., are easily split along bedding planes containing fossils, so collection in such rocks is relatively simple and easy. It is far more difficult to collect fossils from hard, massive rocks like limestones, quartzites and cherts, where fossil remains are often so closely associated with the surrounding rock that a tap from a hammer will break away a fragment of the fossil together with the enclosing rock. In such cases, experienced collectors will rather look for fossils on weathered boulders or on bedding planes, from which they can be extracted much more easily than from the fresh, unweathered rock.

Modern palaeontologists also pay great attention to collections from soft and weathered parts of rocks, especially limestones damaged along tectonic lines, e.g. fractures, faults, etc., by intensive weathering. If such weathered parts, which a palaeontologist of earlier times would have looked upon with extreme distaste, are washed and the remains are sifted on sieves, they will frequently yield a large quantity of excellently preserved and often complete specimens of small fossils, e.g. brachiopods, gastropods and echinoderms, etc., which would never be detected in fresh rocks.

One of the worst mistakes to make, which was common in the early days of palaeontological research and is still occasionally encountered today, is to collect only 'positives' of fossils and not 'negatives', i.e. impression. We should bear in mind that the 'positive' is often no more than the plain filling of the shell or body case and that the true surface, with various important sculptures and even organs, is preserved in the despised and rejected 'negative'. In certain groups of animals, some species and even higher taxonomic units can be determined only from the combined 'positive' and 'negative'. We will give two classic examples. Large numbers of siliceous, originally calcareous, concretions with a splendidly preserved fauna appeared in central Bohemian middle Ordovician slates. The fauna included sparse remains of two representatives of completely different, cystoid echinoderm families with the same pear-shaped theca, but quite differently constructed, taxonomically important organs of the ambulacral (water vascular) system. The first, *Pyrocystites,* had a small mouth, and its very long, narrow ambulacra branched into short secondary canals terminating in kidney-shaped facets for the arms. The representative of the other family, *Archegocystis,* had a large, pentagonal mouth from which five short ambulacra terminating in brachial facets fanned out. These findings were all obtained from only five specimens of which the enlightened collectors kept the impressions, or which were found comparatively recently. The remaining, carefully collected inner moulds, which number about 150, have a nice, pear-like form, but are no use at all, and the expert can only look upon them with glum disappointment.

The other example is even more striking. The carapace (dorsal exoskeleton) of large trilobites of the genus *Selenopeltis,* which lived and swam in middle Ordovician seas, often carried sessile, discoid edrioasteroid echinoderms belonging to the genus *'Hemicystites'.* Sometimes there are over 20 of them on one carapace and we should have difficulty in finding *Selenopeltis* without its complement of tiny guests. This knowledge is only recent, however, and came to light through the careful collection of counter-impressions of the carapace of *Selenopeltis.* The 'positives', which were preserved as internal moulds and were once the only part collected, are completely smooth and show nothing at all.

We are not trying to say that 'positives' are valueless. On the contrary, they often show the impressions of internal structures, such as the imprints of muscles and organs. It is, however, a principle that for complete knowledge of a fossil we must collect both 'positives' and 'negatives', wherever possible.

THE FOSSIL COLLECTOR'S EQUIPMENT

The fossil collector needs only a few simple, but essential, aids for the removal, protection and transport of his specimens. His basic tool is naturally a good steel hammer, or a set of hammers of different sizes, for cracking the rocks and loosening the fossils. Among the various special geological hammers, the best is one with a chisel edge at one end, which can be used like a hoe on friable rocks. Palaeontological or stonecutter's chisels and needles for exposing fossils and prising them out of the rock are further important aids. A magnifying glass or hand lens is also essential; it need not give more than 4- to 10-fold magnification, but it must have the largest possible viewing field. Small fossils lying freely in weathered rocks can be picked out with fine steel forceps. And since we are going prospecting, we need a topographical and geological map, a notebook and a camera. Localities are designated according to their position on the topographical map or their local name. The position of faults can be marked on the map and their appearance, or the positions of fossiliferous layers, can be sketched in the notebook. Even a diagrammatic sketch is better than none at all, and a good photograph is likewise useful. There is no need to be afraid that we shall spend more time at the site with pen and pencil than in actual searching and collecting, however. Properly speaking, every fossil ought to be labelled before it is packed, but it is sufficient if we put finds from the same horizon in the same strong bag and label this with a reference number or a description. Ordinary booklets of coloured and numbered tickets can be used; they can be attached to the fossils, using different colours for different layers or localities, and then all that needs to be done is to enter the relevant field data (e.g. dip, strike of bedding, etc.) together with the appropriate number in the notebook. Bear in mind that the worst thing the collector can do is to rely on so fickle an instrument as human memory. We may be able accurately to identify and classify a fossil a hundred times in succession, but if we once forget where we found it, the loss can never be made good in full. And in that case, even the finest fossils lose their scientific value.

Newly collected fossils need to be packed and transported with the greatest care; abraded and scratched specimens can never be completely restored. Solid specimens are packed — if possible separately — in newspaper, while fragile or small objects should preferably be packed in cotton wool and placed in small boxes or tubes. The whole of the collector's equipment is carried in a strong rucksack or satchel. Large collections to be transported by road or rail are carefully packed in cases. Flat rock fragments containing fossils are always stood upright to minimize the danger of their being broken.

THE PREPARATION AND CONSERVATION OF FOSSILS

Newly found specimens are seldom in such good condition that they can immediately be included in a collection or subjected to palaeontological evaluation. Most of them first need to be cleaned and tidied up; rock fragments must be removed; cracked and broken specimens must be stuck together, and some fragile fossils must be conserved to prevent them from breaking up altogether. For rough work, pincers or a sharp-jawed vice, one or two good chisels and needles and a firm working surface are usually sufficient. A large sack filled with sand, or, better still, a slab of lead makes the best working surface. A good magnifying glass or a binocular microscope is a further essential. After that, only sensitive fingers and patience are needed. Miniature electric drills known as vibrotools, with a set of special needles and chisels, have recently proved very successful in the preparation of fossils. In certain special cases, chemicals can also be used, e.g. to release siliceous shells from limestones ('etching').

Newly found fossils generally need to be washed; only those from unconsolidated, argillaceous shales should preferably be cleaned with spirits and a soft brush, to avoid the water turning a handsome specimen into a shapeless lump of soft clay. Excess pieces of rock are nipped off with pincers or in a vice. We should never try to free a fossil from a rock by force, however, unless it is a free stone core. The outer surface is often attached to the rock, and a blow could easily damage it or smash it to pieces. In any case, the parent rock in which the fossil is preserved can itself tell us a great deal about the stratigraphic position or conditions of the animal's original environment. Cracked or disintegrated fossils should be stuck together as carefully as possible. Adhesives containing acetic acid are unsuitable only for limestones, as in time the acid corrodes calcareous fossils; starch adhesives and other special preparations are satisfactory.

Fragile fossils can be conserved with a solution of acrylite in acetone, for example, and with many other reagents. These must always be highly diluted, of course, and be just strong enough to saturate the fragile walls of the shells, but not so strong that they form, on the surface of the fossil, a solid crystal layer which covers up all the important details and cannot afterwards be removed. The preparation and conservation of microfossils and skeletal remains is a specialized process and it is advisable to consult a special manual.

For the study and exact identification of the species of certain groups of invertebrates such as corals, vryozoans and brachiopods, etc., it is essential to determine the construction and inner structure of their skeletons or shells (if these have been preserved). This is done by the polished and thin section technique.

The polished section technique is the simpler of the two. A piece of rock with a fossil, or an isolated fossil, is sectioned with a carborundum or diamond saw (in the case of soft rocks, a metal saw is sufficient), and the appropriately oriented section surface is ground and polished with fine abrasives (usually various grades of carborundum powder). The resultant polished section can be observed and photographed in incident light, or in immersion fluids which bring out the details, e.g. in alcohol or aniseed oil. Polished sections can also be stained, or treated with acids (etched) to emphasize their relief.

The thin section is a sliver of rock containing a fossil, or part of a fossil, in which the various details are studied under the microscope in transmitted light. As a rule, the appropriately sectioned and polished fossil is mounted in Canada balsam on a glass slide, and the superfluous upper part of the rock or fossil is ground away until the section is sufficiently thin (30 microns, usually) and transparent. If required, it can be stained to differentiate various tissues and structures, or can be covered with a cover slip to obtain a permanent specimen.

THE FOSSIL COLLECTION

The properly cleaned, prepared and conserved fossils, each with a card giving the necessary information, including a reference number, can then be placed in the collection. It is useful if the same number and further information (e.g. the abbreviated name of the locality) can also be written on the specimen itself, somewhere where it is not in the way, such as on the bottom or side of the parent rock. Paper labels are not satisfactory as they may come unstuck. In the case of small fossils kept in boxes or tubes, we should not forget to place a small label inside the container as well.

The organization and arrangement of the collection and the direction taken by the collector's activity depend entirely on the interest and possibilities of the collector himself. There is no fixed ruling or dogma. However, there is one natural law for all collections, although it has never been proclaimed publicly. Every enthusiastic collector soon finds out that his material is getting out of hand. He must therefore decide as soon as possible what he is going to collect and how he is going to organize it, so that it is orderly and he can find his way about his collection without difficulty. Compactness is another important condition; fossils are not postage stamps, and not everybody has a home with enough free space to allow him to fill it ad libitum with his hobby. Unless the collection is on show, it is best to arrange it according to localities, stratigraphic horizons and strata, or classification, or by a combination of both these methods. This allows both the documentation of associations, and hence study from the ecological and biostratigraphic aspect, and biological study of selected groups of animals in which the collector is most interested (Trilobita, Anthozoa, Bivalvia, etc.).

We usually put fossils into boxes, if possible of the same type, but of different sizes, each size having double the area of the size before. This has practical grounds, as it allows us to fill the drawers in our cupboard completely. Otherwise we can make do with cases, which are very useful anyway, either for large locality collections, or for material prepared and stored for classification and study. The cases must naturally have a catalogue number and a suitable label (with an extra one inside) and the contents must be protected from dust.

There is one group of palaeontological material which is generally to be found only in museums, universities and other scientific institutions, but is also accasionally encountered in private collections. I refer to type specimens (holotypes, etc.) from which new species were determined and described. Such type material is exceptionally valuable and must be protected at all costs from damage, destruction and loss. It must likewise be clearly marked to avoid any danger of its being thrown out or confused with something else, and it must be kept separate from other study collections. These are only logical measures when we realize that such specimens are often needed for purposes of revision and by experts from other countries, who, before they can identify some new find of their own, must compare it with the type species or populations. There are, therefore, international agreements which rule that type material must be made available to competent researchers.

THE IDENTIFICATION AND CLASSIFICATION OF FOSSILS

Every serious fossil collector naturally wants to know just what he has found; he therefore tries to identify and classify it and to find out as much as possible about it. When he consults experts or palaeontological literature, he comes to grips for the first time with the problem of classification in palaeontology and hence with the foundations, or basis, of the scientific classification of material.

Every science, including palaeontology, must sum up the results of its study in a clear, understandable form. It is beyond the capacity of the human brain and beyond the possibilities of speech, to give every one of the myriads of objects by which we are surrounded its own individual name. The objects with which man comes into contact must therefore be grouped according to their similarity or their relationships. This need was recognized long ago, and our forebears arranged concepts in a given hierarchy. For instance, the terms horse, ungulate, mammal, animal are the expressions of a classification principle which is actually the hierarchic organization of taxa from the narrowest (horse) to the most general (animal). Modern biological classification, and hence palaeontological classification, is based on the same principle. The basic unit is the species, which is fixed and is not interchangeable (e.g. *Equus przewalskii*, Przewalski's horse). The next highest unit is the genus (*Equus*, horse), which is more general and may comprise one or several similar species. Thus classification progresses to more and more general concepts and also to wider, though clearly definable groups. In a simplified scale, the family (Equidae) is thus above the genus, the order (Perissodactyla, odd-toed ungulates) is above the family, the class (Mammalia) is above the order and the phylum (Chordata) is above the class.

That is all very well, the collector may say, but what are we to do with all those queer foreign terms and why not use vernacular names? Attempts to do precisely this were made in the last century, but they were a failure, for two reasons. Latin or latinized names are international and are thus common ground on which experts from all over the world can meet; in addition, no single language has names for every single living animal, and certainly not for fossils, and where such names do exist they do not generally become established, and their use is extremely restricted. Latin names can be learned and memorized mechanically, and in time their memorization will become quite easily.

The principle of the international scientific classification of animals, including fossils, is very simple. Its author is Carl Linné (Carolus Linnaeus), who introduced the system of binomial nomenclature. The first name is the generic name and the second the name of the species (specific name). For example, the Latin name of Przewalski's horse is *Equus przewalskii*, *Equus* being the generic name. As a further help to the research worker, the scientific name of a species is followed by the name of the author who originally described the species and the year in which this description was published. The same rules apply in palaeontology; as an example we give the complete scientific name of one of the trilobites illustrated in this book, i.e. *Harpides grimmi* Barrande, 1872.

If for any reason the generic name of a species is altered, e.g. through reclassification of the species into another genus or through division of a too broadly conceived, cumulative genus into several allied genera, the name of the author of the species is placed within brackets. For example, the trilobite *Odontochile hausmanni* was originally described by Brongniart as *Trilobites hausmanni*. When the cumulative genus *Trilobites* was divided into separate smaller genera, Brongniart's species was included in the genus *Odontochile*, so that its present scientific name is *Odontochile hausmanni* (Brongniart, 1822).

Latin or latinized terms, often with prescribed endings, are likewise used for higher systematic units; families, for example, usually end in -idae. The formation of the names of orders and higher taxonomic units is not so restricted.

Just one more interesting point. Not so long ago it was suggested that the 'Linnaean' system ought to be revised and its strange names, which are sometimes hard to remember, replaced by a numerical code. This move encountered an insurmountable obstacle, however. It was found that nature had not been very generous towards some palaeontologists when giving out mathematical brain cells, and unfortunate scholars who had difficulty in remembering their own telephone number froze in horror at the thought of having to remember that the trilobite *Paradoxides gracilis* would in future be known as number 281735 42218. When it was realized that the proposed numerical system would not produce the desired effect without extensive recourse to computers, it fell into disuse, and the tried and proven codified Linnaean classification is still the official system.

INDEX OF SCIENTIFIC NAMES

Acantherpestes 138
Acantherpestes vicinus 138
Acanthochitona calliozena 48
Acanthodii 178, 179
Aegiromena aquila 40
Agetochorista tillyardi 137
Agnatha 174, 175
Agnostida 102
Agraulos ceticephalus 106
Aistopoda 194
Ammodiscus 17
Amphibia 186—195
Anaconularia anomala 22
Anahamulina subcylindrica 85
Anarcestes plebejus 76
Anarcestina 76
Anchordata 168
Anthozoa 24—29
Anthracosauria 188
Anthracosaurus 188
Anthracosaurus dyscriton 188
Antiarchi 176
Arachnocystites infaustus 140
Archaeogastropoda 58
Archegocystis 214
Archegonaster 160
Archegonaster pentagonus 160
Archiscudderia 139
Archiscudderia tapeta 139
Aristocystites bohemicus 142, 143
Aristozoe 93
Aristozoe memoranda 92, 93
Arthodira 177
Arthroacantha 203
Arthropoda 90—139
Articulata 36—47
Asaphacea 108
Asteriacites fallax 204
Asteroidea 164, 165
Asteropyge punctata 129
Asteropygidae 129
Astylospongia 19
Astylospongia praemorsa 19
Atrypa 46
'*Atrypa*' *renitens* 46
'*Atrypa*' *verneuiliana* 46
Atrypidina 46
Aulacopleura konincki 118
Aviculopecten multiplicans 56

'*Barrandeoceras*' *bohemicum* 67, 74
Bathmoceras praeposternum 68
Batostoma 30
Batostoma minnesotense 30
Batostoma poctai 30
Bellerophon 58
Bellerophon vasulites 58
Beyrichia 95
Beyrichia latispinosa 95
Beyrichia moodeyi 95
Billingsella 37
Billingsella coloradoensis 37
Birkenia 175
Bivalvia 52—57
Blastoidea 144—147
Bohemoharpes ungula 121
Bohemopyge 109
Bohemopyge discreta 108, 109
Boiotremus 59
Boiotremus fortis 58, 59
Bojoptera colorata 135
Bojoscutellum paliferum 113
Bolloceras 73
Bolloceras rex 73
Bothryolepis 176
Bothryolepis canadensis 176
Brachiopoda 34—47
Branchiosaurus 186, 187
Branchiosaurus salamandroides 186
Bryozoa 30—33
Bumastus 114
Bumastus hornyi 114

Calamoichthyes 180
Calcarea 18
Calceola 25
Calceola sandalina 25
Calcichordata 158, 172, 173
Calcispongea 18
Caleidocrinus multiramus 152
Calymenina 124
Camerata 150
Cancrinella altissima 42
Cardiola 53
Cardiola docens 52, 53
Carneyella pilea 203
Cephalopoda 64—85
Ceratiocaris 93
Ceratiocaris bohemicus 93
Ceratiocaris papilio 93

Ceratites 80
Ceratites semipartitus 80
Cheirocrinus 140
Cheirocrinus insignis 140
Cheirolepis 181
Cheirurina 124
Cheirurus 124
Cheirurus insignis 124
Chelodes 49
Chelodes bohemicus 49
Chonetes tardus 42
Chonetidae 42
Chordata 168, 174—199
Climatius 179
Clymenia 78, 79
Coelenterata 20—29
Collignoniceras woollgari 84
Conchidium 39
Condylopyge 102
Condylopyge rex 102
Conocoryphe 104
Conocoryphe sulzeri 104
Conularia 203
Conularia niagarensis 22
Conulata 22
Corbuloceras corbulatum 70
Cornuproetus 117
Cornuproetus venustus 116, 117
Cornuta 173
Cothurnocystis elizae 173
Crinoidea 150—155
Crossopterygii 184, 185
Crotalocrinites 152
Crotalocrinites rugosus 152
Crustacea 92—95
Cryphops cryptophthalmus 127
Cryptodonta 52, 53
Cryptostomata 32—33
Ctenocephalus coronatus 105
Ctenodonta 54
Ctenodonta bohemica 54
Ctenodus obliquus 182, 183
Cyclomedusa 20
Cyclomedusa plana 20
Cyclopygacea 110
Cymostrophia stefani 41
Cyrtia 44
Cyrtia exporrecta 44
'*Cyrtoceras*' 68
Cystoidea 140—143

Dalmanitidae 128
Dalmanitina proeva 129

221

Deltoblastus timorensis 147
Dendrocystites
  barrandei 159
Dendroidea 168, 169
Dicranopeltis scabra 130
Dicranurus monstrosus 133
Dictyonema
  flabelliforme 169
Dimorphichnus obliquus 204
Dinichthys
  (Dunkleosteus) 177
Diplopoda 138, 139
Diploporita 142
Diplovertebron 189
Diplovertebron
  punctatum 189
Dipnoi 182
Discosauriscus 191
Discosauriscus
  potamites 190, 191
Discosauriscus
  pulcherrimus 191
Dolichosoma
  longissimum 194
Drahomíra 51
Drevermannia 117

Echinodermata 140—167
Echinoidea 166, 167
Ectillaenus parabolinus 114
Edaphosaurus sp. 196, 197
Edrioaster bigsbyi 157
Edrioasteroidea 156, 157
Elasmobranchii 178—179
Ellesmeroceratida 68
Ellipsocephalacea 100
Ellipsocephalus 101
Ellipsocephalus hoffi 100, 101
Ellipsostrenua gripi 101
Ellipsotaphrus 111
Ellipsotaphrus
  monophthalmus 111
Encrinaster 163
Encrinaster roemeri 162, 163
Endoceras sp. 64
Endoceras novator 65
Endoceratoidea 64
Entelophyllum 24
Entelophyllum
  prosperum 24
Eochelodes 49
Eocrinoidea 148, 149
Eoharpes benignensis 121
Eoporpita 21
Eoporpita medusa 21
Eospirifer togatus 44
Equus przewalskii 219
Ettoblatina 137

Ettoblatina bohemica 137
Eucephalaspis 175
Eucephalaspis lyelli 174, 175
Euryalae 162
Eurypterida 90, 91
Eurypterus 91
Eurypterus fischeri 90, 91
Eusthenopteron 185

Famatinolithus noticus 123
Favosites 27
Favosites gothlandicus 27
Favosites tachlowitzensis 26, 27
Fenestella antiqua 32
Fenestrellina 32
Fenestrellina capilosa 32
Fissiculata 145
Flexibilia 154
Flexicalymene 124
Flexicalymene incerta 124
Foraminiferida 16, 17
Furca bohemica 96
Furcaster 163
Furcaster palaeozoicus 163
Fusulina sp. 16

Gastropoda 58—63
Germaropyge 101
Germaropyge germari 101
Gogia 149
Gogia prolifica 149
Gompholepis panderi 183
Gonioclymenia speciosa 78, 79
Graptoloidea 170, 171
Gymnites 83
Gymnites incultus 83
Gypidula 39
Gypidula caduca 38
Gyrocystis barrandei 159

Halysites catenularia 27
Harperopsis 94
Harperopsis bohemica 94
Harpides 120
Harpides grimmi 120, 219
Harpina 120
Helcionella 58
Helcionella subrugosa 58
Heliolitoidea 28
Heliolites 29
Heliolites decipiens 28, 29
Helioplasma 29
Helioplasma aliena 29
Helioplasma kolihai 29
Heliotidae 28
Helminthopsis 205
Hemichordata 168—171

Hemicystites 156, 214
Hernodia 203
Heteroschima gracile 145
Hexameroceras 72
Hexameroceras panderi 72
Homalozoa 158, 159
Homocystites alter 140
Hyalospongea 18, 19
Hydnoceras 19
Hydnoceras tuberosum 18, 19
Hydrocephalus 99
Hydrocephalus carens 99
Hydrozoa 20, 21
Hyolitha 88, 89
Hysterolites 44
Hysterolites nerei 44

Illaenidae 114
Inadunata 152
Inarticulata 34, 35
Insecta 134—137
Intrapora irregularis 203
Ivdelinia 39
Ivdelinia procerula 39

Jivinella 37
Jivinella incola 36, 37
Junocrinus globulus 201

Keraterpeton 192
Keraterpeton crassus 193
Keriophyllum tabulatum 25
Koremagraptus 169
Koremagraptus
  spectabilis 168, 169

Labyrithodontia 186
Lagynocystites
  pyramidalis 173
Latimeria 185
Latimeria chalumnae 184, 185
Lechritrochoceras 71
Lechritrochoceras
  trochoides 71
Leiopteria 56
Leiopteria mucro 56
Lepidocentrus 167
Lepospondyli 192
Leptaena 40
Leptaena depressa 40
Leptostraca 92
Lichenoides priscus 148
Lichida 130
Lingula 34, 35
Lingulella 34, 35
Lingulella insons 34
Lingulobolus 34

Lingulobolus feistmanteli 34
Loxonema sinuosum 63
Lycosuchus sp. 198, 199

Machaeracanthus
　bohemicus 179
Macroscaphites yvani 85
Mantelliceras mantelli 83
Marella 96
Marella splendens 96
Maxilites maximus 88, 89
Megalaspides
　dalecarlicus 108
Meganeura sp. 134
Megistapsis alienus 109
Michelinoceratida 68
Microparia 110
Microphaga 72
Mimagoniatites 77
Mimagoniatites fecundus 77
Mitrata 172—173
Mitrocystites mitra 172
Mollusca 48—85
Mollusca (?) 86—89
Monograptus lobiferus 171
Monograptus priodon 170, 171
Monoplacophora 50, 51
Monotrypa kettneri 31
Monotrypa undulata 30
Myzostomida 201

Nanillaenus 115
Nantiloidea 66
Nautilus 66, 74
Nautilus pompilius 74
Neoceratodus 182, 183
Neoceratodus forsteri 183
Neoloricata 48
Neopilina 50
Neopilina galatheae 50
Neoptera 136
Nisusia kuthani 37
Novakella 110
Novakella bergeroni 110
Nowakia 87
Nowakia cancellata 86
Nuculana 55
Nuculana pernula 55

Obolus 35
Obolus complexus 35
Octamerella
　calistomoides 72
Odontochile 128, 220
Odontochile hausmanni 128, 129, 219, 220
Odontopleura ovata 133
Odontopleurida 132

Ophiderpeton 194
Ophioceras 67, 71
Ophioceras rudens 71
Ophiurida 162
Ophiuroidea 162, 163
Opsimasaphus nobilis 108
Oriostoma 62
Oriostoma eximium 62
Orthacanthus bohemicus 179
Orthacea 36
Orthis 37
Orthis callactis 37
'Orthoceras' 68
'Orthoceras' arion 66, 67
Orthonychia 61
Osteichthyes 180—185
Ostracoda 94
Ostrava 134
Ostrava nigra 134
Ostrea 56
Otarion diffractum 118

Palaeoconcha 52, 53
Palaeoloricata 48
Palaeonisciformes 180
Palaeotaxodonta 54
Paleoxyris 179
Panenka 53
Panenka expansa 53
Paradoxidacea 98
Paradoxides gracilis 98, 99, 219
Parakionoceras 69
Parakionoceras originale 69
Paramblypterus rohani 180, 181
Parapyxion 95
Parapyxion subovatum 95
Pauxilites 89
Pauxilites solitarius 89
Peismoceras 67
Pelycoscuria 196
Pentremites 146
Pentremites godoni 146
Pernerocrinus 153
Pernerocrinus paradoxus 153
Peronopsis 103
Phacopida 124
Phacopidae 126
Phacops rana 126
Phacops rana
　crassituberculata 126
Phalagnostus nudus 102
Phestia 54
Phestia attenulata 54
Phillipsia 119
Phillipsia gemmifera 119
Phthinosuchus 199
Phyloblatta 137

Phyloblatta sp. 136
Phyllocarida 92
Pinacites jugleri 77
Placodermi 176, 177
Placoparia barrandei 125
Platyceras 60
Platyceras anquis 61
Platyceras elegans 61
Platyceras gregarium 60, 61
Pleurojulus 139
Pleurojulus levis 139
Polydeltoideus plasovae 144, 145
Polydora 207
Polyplacophora 48, 49
Polypterus bichir 180
Porifera 18, 19
Porpita 21
Posidonia 57
Posidonia becheri 57
Praecardium 53
Pricyclopyge binodosa 110
Productidae 42
Proetacea 116, 118
Proplacenticeras 83
Proplacenticeras
　orbignyanum 82, 83
Protaxocrinus 155
Protaxocrinus svobodai 155
Proteocystites flavus 143
Protopalaeaster 164
Protopalaeaster
　narrawayi 164
Protorthoptera 137
Protospongia mononema 19
Protospongia novaki 19
Protozoa 16
Ptenoceras 67
Pteraspis 175
Pteraspis cornutus 175
Pterygotus 91
Pterygotus bohemicus 91
Pterygotus buffaloensis 91
Ptychoparia 104
Ptychoparia striata 104
Ptychopariina 104
Pycnosaccus 155
Pycnosaccus bucephalus 155
Pyritocephalus sculptus 181
Pyrocystites 214

Radotina kosorensis 177
Reedops cephalotes 126
Reptilia 196—199
Reteporina prisca 32
Retipilina knighti 50
Rhabdammina 16
Rhenechinus 167

Rhenechinus hopstätteri 166, 167
Rhipidomena 203
Rhizocorallium 207
Rhizopodea 16, 17
Rhombifera 140
Richthofenia 42
Richthofenia lawrenciana 42
Rizosceras 67
Rugosa 24, 26

Sao hirsuta 106
Scaphites 84, 85
Scutellum 201
Scyphocrinites 61, 151
Scyphocrinites elegans 150, 151
Scyphozoa 22, 23
Selenopeltis 133, 156, 214
Selenopeltis buchi 132, 133
Seymouria 190, 191
Seymouriamorpha 190
Sieberella 39
Sieberella sieberi 38, 39
Silicea 18, 19
Siluraster 164
Siluraster perfectus 164
Solenopleura canaliculata 107
Solenopleuracea 106
Somasteroidea 160, 161
Sphaerexochus mirus 124
Sphaerocodium 202
Sphenothallus 23
Sphenothallus angustifolius 23

Spiniscutellum umbelliferum 112
Spiraculata 146
Spiriferida 44
Stenopareia 114
Sthenaropoda fischeri 137
Strenuella 101
Striatostyliolina 87
Striatostyliolina peneaui 87
Stringocephalus 47
Stringocephalus burtini 46
Stromatocystites 156
Stromatocystites pentangularis 156
Strophomenidina 40
Styliolina 87
Styliolina fissurella 86

Tabulata 26
Taxocrinus 154
Taxocrinus coletti 154
Tentaculites 86
Tentaculitida 86—87
Terataspis grandis 130
Terebratulidina 46
Thallograptus 169
Thallograptus muscosus 169
Therapsida 198
Thysanopeltidae 112
Thysanopeltis 113
Thysanopeltis speciosum 113
Tigillites 207
Tigillites vertebralis 207
Tirolites 81
Tirolites idrianus 81
Tremanotus 59

Tremanotus civis 59
Trepostomata 30, 31, 32
Tretaspis 122
Trilobita 98—133
Trilobites hausmanni 220
Trilobitoidea 96
Trilobitomorpha 96, 97
Trimerocephalus 127
Trinucleina 122
Trinucleoides reussi 122
'Trochoceras' 68
Trochocystites 159
Trochocystites bohemicus 158, 159
Troostocrinus reinwardti 146
Tryblidium 51
Tryblidium reticulatum 51
Tubina 62
Tubina armata 63
Turrilites costatus 85

Urocordylus scalaris 192

Villebrunaster thorali 160
Vioa 201

Warburgella 117
Weberides 119
Weberides mucronatus 119
Wocklumeria 79
Wocklumeria sphaeroides 79

Xenacanthus 179
Xenacanthus bohemicus 178, 179
Xenodiscus plicatus 81
Xylodes 24